W9-AXL-050

Digital Computer Control Systems

Digital Computer Control Systems

G. S. Virk

Department of Control Engineering
University of Sheffield

McGraw-Hill, Inc.

New York St. Louis San Francisco Bogotá Caracas
Mexico Montreal San Juan São Paulo Toronto

1 2 3 4 5 6 7 8 9 0 DOC/DOC 9 8 7 6 5 4 3 2 1

ISBN 0-07-067512-0

First published 1991 by MACMILLAN EDUCATION LTD.
Houndmills, Basingstoke, hampshire RG21 2XS
and London.

Printed and bound by R.R. Donnelley & Sons Company.

To Ann, Bernadette and Térese,

and to the memory of my Father.

Contents

Preface

Digital control methods are replacing most conventional analogue systems. It is therefore important that the control engineers of the present and future are familiar with the features and facilities available, and are aware of the limitations of such systems. The aim of the book is to present an up-to-date introduction to these systems, and to instruct and assist readers so that they are able to connect digital computers to real systems for control and/or analysis purposes. The main aspects such as discretisation and system design methodologies which need to be considered for solving a particular application are discussed so that the reader is able to successfully perform a complete system design for his/her problem.

The material assumes that readers have completed a first course in control systems where the classical continuous domain methods have been covered, as well as some of the concepts of the state-space approach. Hence knowledge of such areas as the Laplace transform, frequency (Bode, Nyquist, Nichol, etc.), and time domain analysis methods, root locus, compensator design, etc., is assumed. These methods are extended to the discrete z-domain so that computer control can be implemented. Both input/output (classical control) and state-space (modern control) methods are presented with application to illustrative examples. At the end of some of the chapters, numerical problems are included with sample solutions provided in an appendix.

It is not the intention to concentrate on one particular application area but to present an overall systems design approach, so that the generic control methods can be applied to a variety of different areas. Hence the book should be of interest to third year undergraduate or master's students in all branches of engineering and applied mathematics as well as practising engineers and scientists.

The book chapters are set out as follows. The first chapter gives a historical perspective on computer control, and considers aspects such as interfacing (A/D and D/A converters), sampling rates, quantisation effects, hold devices and the problems caused by aliasing. In chapter 2 the z-transform and its properties are introduced, and how it is used in studying digital processes and systems. Discrete approximations can simplify the analysis greatly and are therefore considered in some detail. Since the z-transform analysis can only define the time responses at the discrete sampling instants, the modified z-transform is also presented so that inter-sampling time values can be described. In addition, the usefulness of multi-rate sampling is discussed so that computer resources can be optimised and appropriate sample rates used to maximise the effectiveness.

Chapter 3 considers the study of systems in the z-domain. The stability methods of Routh, Jury and Raible for sampled-data systems are presented and how time domain, frequency domain and root locus analysis are performed. Most of the methods are extensions of the well known continuous domain techniques. Controller design is the main objective of chapter 4, where several different designs for a rigid-body satellite example are presented. The methods illustrated include continuous and digital domain analysis (in the w'-plane); root locus design in the z-domain; state feedback control; digital PID implementation; deadbeat response design; and optimal controllers. All the designs are performed on the same example so that the different controllers can be compared objectively. A brief introduction to sampled-data control systems under noisy conditions is also presented.

Chapter 5 considers the overall hardware and software considerations in real-time computer systems and how such systems can be designed. The important features such as hardware requirements and software aspects that control engineers need to be aware of when constructing a suitable system are presented. Guidance on the co-ordination of the overall system design is also given. In chapter 6 some of the latest developments, namely artificial intelligence, and parallel processing, are discussed. These are likely to have an enormous impact on control applications where the computers will use concepts such as fuzzy logic and other knowledge-based expert systems with appropriate inference engines to think and deduce solutions in a similar fashion as humans do. Such computers will need to possess vast processing power that can only realistically be achieved using parallel processing techniques. Here the computational task can be distributed over a network of processors that can work on sub-tasks and can communicate between each other so that the overall solution can be determined. The majority of computer users at present are experienced only in the traditional sequential "flow diagram" programming approach. The effective programming of parallel computers requires the users to be retrained so that problems such as deadlock, problem partitioning, communication synchronization, computer architecture optimisation and program termination can be solved. These issues are discussed with particular emphasis to transputer systems programmed in occam.

Much of the work described in this book has arisen from lecture courses given by the author to students in the Department of Control Engineering, University of Sheffield as well as to students at Sheffield City Polytechnic and University of Southampton. Constructive comments received from undergraduate and postgraduate students over the years have improved the material greatly. Also many colleagues have assisted and commented on earlier drafts of some of the chapters. This feedback is gratefully acknowledged and appreciated. In particular the author is grateful to Ian Durkacz, Panos Kourmoulis and Gurmej Virk for their help in preparing sections of the manuscript and diagrams. Dennis Loveday and Joe Cheung proof-read the whole script and made very useful suggestions. In addition the comments made by the series editor Paul Lynn on an earlier draft of the book were invaluable and have been gratefully received to improve the presentation of the manuscript. Finally the author wishes to pay tribute to his wife Ann Marie for her encouragement and understanding over the years. Without her moral and physical assistance this book would never have been accomplished.

Gurvinder Singh Virk
January 1991

Digital Computer Control Systems

1 Computers in Control Systems

1.1 Introduction

In its relatively short existence, digital computer technology has touched, and had a profound effect upon, many areas of life. Its enormous success is due largely to the flexibility and reliability that computer systems offer to potential users. This, coupled with the ability to handle and manipulate vast amounts of data quickly, efficiently and repeatedly, has made computers extremely useful in many varied applications. The ability to communicate rapidly between remote networks has led to their extensive use in banking and commerce, because transactions can be conducted through links such as fibre optic cables or telephone channels. In fact the situation has advanced to the stage where the transactions can be initiated and completed without any human interaction.

Our domestic lifestyles, also, have been affected by the inclusion of digital processors into household appliances, thereby making them semi- or fully-automatic. For example, we can now program ovens to commence cooking at the right time so that the meal is ready when we reach home, video recorders can be set up so that our favourite television programmes are recorded if we happen to be out at the time, and appliances can be programmed to switch on or off at appropriate times automatically. Such devices have given us greater flexibility in, and control over, our lives.

In our work environment, computers have also made major impacts: typewriters are being replaced by word-processing units; mundane shop floor jobs are being performed by machines; finished goods are tested automatically, and so on. The list, already long, continues to lengthen – nowadays with even greater rapidity.

In this book we will concentrate on the role of the digital computer in control systems. Here the computer acts as the controller and provides the enabling technology that allows us to design and implement the overall system so that satisfactory performance is obtained. Digital control systems differ from continuous ones in that the computer can only act at instants of time rather than continuously. This is because a computer can only execute one operation at a time, and so the overall algorithm proceeds in a sequential manner. Hence, taking measurements from the system and processing them to compute an actuating signal, which is then applied to the system, is a standard procedure in a typical control application. Having applied a control action, the computer collects the next set of measurements and repeats the complete iteration in an endless loop. The maximum frequency of control update is defined by the time taken to complete one cycle of the loop. This is obviously dependent upon the complexity of the control task and the capabilities of the hardware.

At first glance we appear to have a poorly matched situation, where a digital computer is attempting to control a continuous system by applying impulsive signals to it every now and then; with this viewpoint it seems unlikely that satisfactory results are possible. Fortunately the setup is not as awkward as it first appears – if we take into account the cycle iteration speed of the computer and the dynamics of the system, we can expect adequate performance when the former is much faster than the latter. Indeed, digital controllers have been used to give results as good as, and even better than, analogue controllers in numerous cases, with the added feature that the control strategies can be varied by simply reprogramming the computer instead of having to change the hardware. In addition, analogue controllers are susceptible to ageing and drifting which in turn cause degradation in performance. Advantages like this have attracted many users to adopt digital technology in preference to conventional methods, and consequently application areas are growing rapidly. Computer control has already been applied to many areas, a few examples of current interest being

(i) autopilots for aeroplanes/missiles,
(ii) satellite attitude control,
(iii) industrial and process control,
(iv) robotics,
(v) navigational systems and radar, and
(vi) building energy management and control.

With advances in VLSI (very large scale integration) and denser packing capabilities, faster integrated circuits can be manufactured which result in quicker and more powerful computers. Application to control areas which a few years ago were considered to be impractical or impossible because of computer limitations, are now entering the realms of possibility. For example, it is now possible for computers to control "intelligent" robots with automatic vision systems in real-time. Although further major research is still needed in such areas, the techniques have been demonstrated, and with further advances in computer technology fully developed solutions are not too distant.

Another recent advance in computer systems is in the area of parallel processing, where the computational task is shared out between several processors which can communicate with each other in an efficient manner. Individual processors can solve subproblems, with the results brought together in some ordered way, to arrive at the solution to the overall problem. Since many processors can be incorporated to execute the computations, it is possible to solve large and complex problems quickly and efficiently. We shall consider these futuristic aspects in later chapters, but for now we start by looking briefly at the historical development of computer control.

1.2 Historical Development

Digital computers have been continuously developed and improved since the mid 1940s but it was not until the early 1960s that they had evolved sufficiently to be used as controllers. Early applications used a central computer which was a large and expensive machine with limited processing power. To ensure cost effectiveness, each machine had to perform several operations in a time sharing manner, and Figure 1.1 shows a typical network.

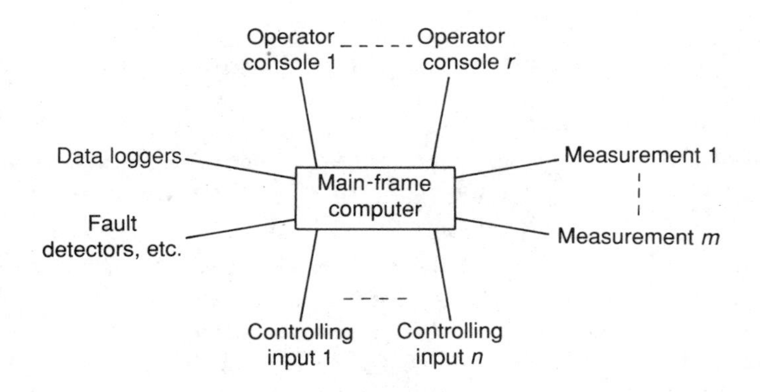

Figure 1.1 Star configuration

Such a philosophy of centralised control led to several disadvantages, mainly because of the limitations and reliability aspects of the computer. Also, the adoption of a star configuration, as it is termed, could lead to a conflict of interest at local and central levels of control. For example, a situation may arise where, despite a particular process requiring its control loop to be updated every second, the central computer can only sample every three seconds because of its global commitments. Discrepancies of this kind lead to poor performance.

Despite such difficulties, interest in computer control grew rapidly. The new technology was seen as an exciting tool that was ripe for development; applications, although primitive by modern standards, increased dramatically. These applications took one of two forms; either the computer provided instructions for human operators to implement, or it changed the set point of analogue regulators. Both forms used classical analogue control equipment.

In the early 1960s a new digital form of interfacing instrumentation was developed so that the computer could measure and control the system directly. This led to the terminology Direct Digital Control (DDC), to emphasise the direct interaction between the computer and the plant under control. DDC allowed different control laws to be implemented by merely changing the software, rather than having to rewire or modify analogue hardware systems. Also, complex interactions in multivariable systems could be easily implemented. The resulting increase in flexibility attracted great interest and accelerated the transfer from analogue to digital technology. Specialised DDC languages evolved, allowing users to enter desired set-points and controller parameters as for conventional regulating control systems, but without the need for specialised programming skills. DDC caught on quickly and, during the years 1963-65, considerable progress was made in algorithm design, choice of sampling rate, control algorithms, reliability and systems analysis.

Another interesting lesson learned from this early work was the need to respond quickly to demands from the process. This led to the addition of a hardware device into the computer which allowed an event in the process to interrupt the computer during the performance of its normal program so that a more urgent task could be undertaken in, for example, an emergency situation (see chapter 5 for further discussion).

Although DDC involved analysis in the digital domain, the computers were still large machines used for performing many tasks. Typically, one computer measured and controlled several hundred different operations. This centralised controlling policy was to change during the 1970s when mini and microcomputers emerged. In contrast to the large main-frames, the latter were designed for small dedicated tasks and were inappropriate for controlling multiple loops. Computer systems of this nature have led to localised control methods in which each computing device performs a specialised task and gives rise to a local control loop in the overall system structure. Individual microprocessors can be connected to each other via a data bus that allows communication to take place. An extension of such a localised control strategy is to incorporate different levels of control action, as in the management structure of a commercial company. A central computer (the managing director) can be linked via a data bus to several other computers (the managers). Further lines of command can radiate to other buses which connect to devices lower down the command level until the processors (the workers) actually performing the control action are reached.

Since the early applications, computer technology has been developing at an ever increasing rate. Hardware is becoming faster and more reliable and software development is increasing correspondingly. Operating systems and high level programming languages have been introduced to allow ease and flexibility in creating specialised code. There are now real-time operating systems and languages specially tailored to real-time control applications.

The rapid development in the application of computer hardware to control problems exposed several shortcomings in other areas. For example, the systems under consideration needed to be understood more precisely for better control – hence better modelling techniques were developed. In fact, extensive research was pursued on the theoretical aspects in order that the applications could be supported. One has only to glance through the journals and conference proceedings covering the years 1960-70 to appreciate the level of effort that was applied to control theory. This is not to say that research ceased after these years. Indeed, the high levels of effort continued well into the 1980s. The situation has been reached where the theoretical design and analysis methods are now so far advanced that one needs to be an applied mathematician to understand the results. Computer control engineers often lack the formal mathematical training and have little chance of comprehending, let alone applying, these results. Consequently most of the theoretical results have not been transferred to applications as yet, and a rift has formed between control theoreticians and practising engineers. The problem has been noticed by both sides, and attempts are being made to alleviate the technical difficulties. There is still some way to go before suitable solutions are obtained and we can only hope they will come soon.

We end this section by mentioning some of the latest developments in computer systems. Two of the main areas of current investigation are artificial intelligence and parallel processing. Both areas are included in the Japanese "Fifth Generation" computer initiative, and will no doubt have an important effect on computer control applications in the near future. For this reason we discuss these aspects later on in the book. At this point we merely note that, with powerful future computers, we shall be able to reconsider applying computer control methods to problems for which such solutions are presently beyond reach. For example, the real-time control of complex systems with fast dynamics should be possible with this new breed of computer. The extra power available will also allow better user-friendly display routines to be driven. Colour graphics that can be updated rapidly will allow real-time simulation, so that operators can see precisely the behaviour of the controlled plant. Much research work in this area is currently underway.

1.3 Digital Computer Interfacing

Having given a brief overview of the evolution of computer control, we can now concentrate on the main problems that need to be confronted in all applications. They can be broadly divided into the areas of interfacing and of design/analysis. We concentrate on the problem of interfacing computers to continuous systems so that the analogue plant signals can firstly be read into the computer, and then digital control signals can be applied to the system. Other forms of interface units are also necessary, for example digital signals from the plant may need to be inserted into the computer (see Bennett [13] and Mellichamp [83] for further discussion). Here we focus on the analogue/digital junctions where, to accomplish the interfacing successfully, a suitable device must be inserted at all input and output ports of the computer. Analogue signals must be converted into digital form for analysis in the computer, and the digital signals from the computer have to be converted back to analogue form for application to the plant under control.

Structure of a Digital Control Loop

In a typical control system, several control loops are present. Each is designed to perform some particular task – for example, when controlling the flight path of an aircraft we may have separate control loops set up to maintain heading, velocity and height. The following procedure must take place in a digital control loop:

Step 1: Measure system output and compare with the desired value to give an error.

Step 2: Use the error, via a control law, to compute an actuating signal.

Step 3: Apply this corrective input to the system.

Step 4: Wait for the next sampling instant.

Step 5: Go to Step 1.

The loop structure which achieves this can take many forms, but two of the most common arrangements are shown in Figure 1.2. The configurations are quite distinct because of the different location of the analogue-to-digital (A/D) converter in the feedback loop. In both cases the conversion is necessary because digital computers can only resolve the continuous data down to a minimum finite level, dictated by the digital word length used. For example, a 2 bit word can distinguish only 4 discrete levels on the full scale as indicated in Figure 1.4. For a given word length, the size of the full scale dictates the magnitude of the quantisation errors in the digitisation process. Since the error channel invariably involves smaller signal levels, digitisation at this point, using smaller full scale values, will produce smaller quantisation errors. Consequently, smoother control will generally be possible compared with the case where the higher level output signal is converted and compared with a reference, as shown in Figure 1.2(b).

Nevertheless, the reference signal is often available digitally – perhaps from some command and control computer – and hence one is forced to

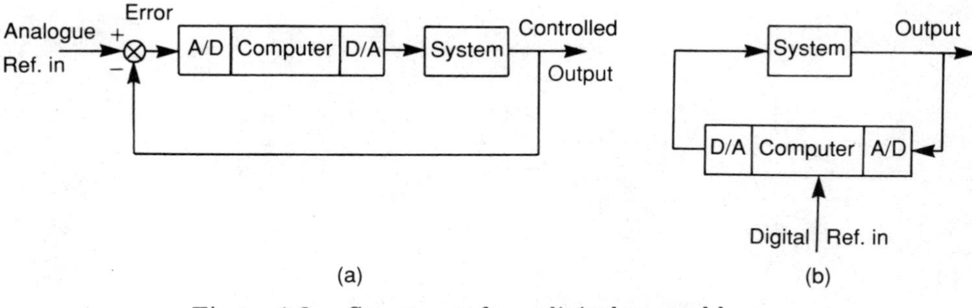

Figure 1.2 Structures for a digital control loop

the configuration of Figure 1.2(b). Typically 10/12 bit conversion levels are used, giving a least significant bit (LSB) as 0.1% of full scale values. Although this seems more than acceptable, digital controllers also involve discretisation in time. Modern microprocessors can easily perform one million operations per second, but a complex control algorithm may require 10^5 operations to complete a cycle. For this algorithm it follows that a cycle rate of 10 Hz is possible. For controlling systems with slow dynamics this will be indistinguishable from the performance given by a continuous controller, but for systems that respond at higher frequencies we may encounter problems in achieving control. The reason is fairly apparent in that we cannot expect to control a system that is changing rapidly when the computer is cycling at a slow rate – this is analogous to asking a slow runner to catch a faster one. Therefore the cycle speed, or the sampling period, plays an important role in the design of the digital controller, as is further discussed below.

Interface Units

As mentioned above, a communication mismatch problem exists between the (digital) controlling computer and the (analogue) system being controlled. In this section we look at a few interfacing units that are used to overcome such problems at various digital computer/analogue system junctions.

Digital-to-Analogue Converter

D/A converters take many forms, but essentially they all rely on the basic idea of using an input digital code to open or close switches in an electronic circuit. The closure of the switches causes a voltage to be generated corresponding to the digital code. A standard circuit is shown in Figure 1.3.

The digital number to be converted is applied to the switches S_1 to S_n, such that "1" causes closure of the switch and "0" opens it. Switch S_1 is the

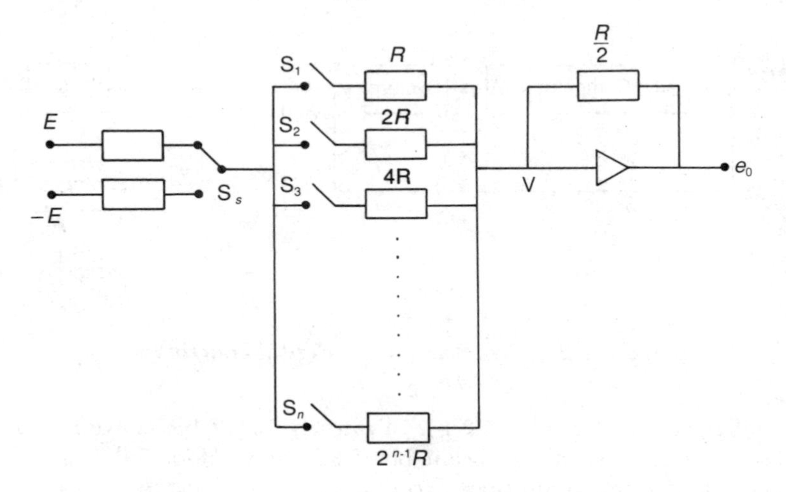

Figure 1.3 Weighted register D/A converter

most significant bit (MSB) and S_n is the least significant bit (LSB), for a word with n bits. An explanation of the circuit can be found in most books on operational amplifier (op-amp) circuits, but we give a brief description for completeness.

Because of the large gain of the amplifier, the point V on the circuit is a virtual earth, and so no current is drawn by the op-amp. The total current flowing from the input resistor branch to the point V must therefore flow through the feedback resistor branch, causing the appropriate output voltage to be generated. For example, if S_1 is the only switch closed, the input current is $\pm E/R$. This flows through the feedback resistor causing an output voltage $\mp E/2$ as required. Similarly, closure of switch S_2 on its own, causes e_0 to be $\mp E/4$. In this way it is straightforward to show that the total output is in accordance with the binary representation of the code, thus

$$e_0 = \mp E\frac{R}{2}\left\{\frac{b_1}{R} + \frac{b_2}{2R} + \frac{b_3}{4R} + \cdots + \frac{b_n}{2^{n-1}R}\right\} \tag{1.1}$$

for a digital word $b_1 b_2 b_3 \cdots b_n$, applied to the input switches. The sign bit S_s, sets the polarity of the output.

Analogue-to-Digital Converter

The conversion of signals from analogue to digital form is a little more involved. To illustrate the procedure, consider the circuit shown in Figure 1.4, where three comparators are required to determine into which of four possible intervals the analogue input falls. A logic network is required to translate the comparator output levels into signals which set the output registers. Since all the comparators work in parallel, fast conversion times are achieved by using these so-called simultaneous A/D devices. The price paid for the speed is that the size and complexity of the circuit grows exponentially with the length of the digital word – an n-bit word requires 2^{n-1} operational amplifiers. For reasonable accuracies, word lengths of around 12 bits are required, corresponding to over 4000 comparators. In addition, a further complication is encountered when using such excessive numbers of comparators – each op-amp needs to have its threshold voltage set with the required accuracy for the circuit to work properly. Clearly, the resolution becomes finer and more difficult to achieve as longer word lengths are used, and a stage is reached where the op-amps cannot be set up as required.

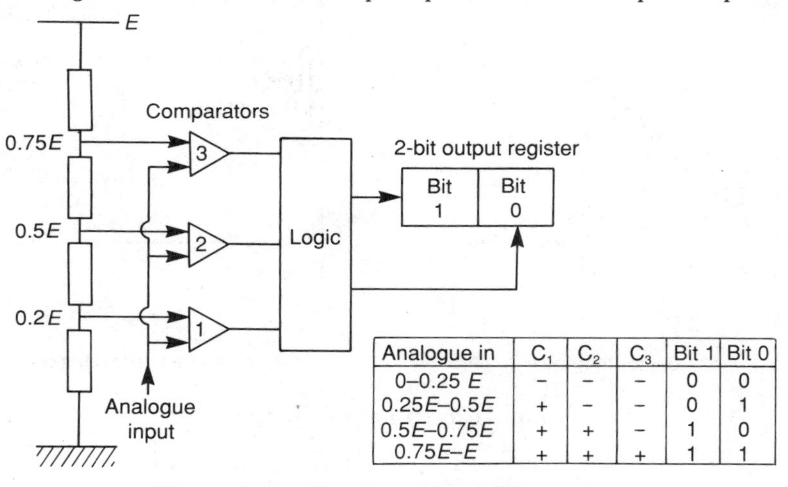

Analogue in	C_1	C_2	C_3	Bit 1	Bit 0
0–0.25 E	–	–	–	0	0
0.25E–0.5E	+	–	–	0	1
0.5E–0.75E	+	+	–	1	0
0.75E–E	+	+	+	1	1

Figure 1.4 Simultaneous A/D converter

A more satisfactory solution to the A/D conversion problem is obtained with a closed-loop method, where feedback around a D/A converter is used, as shown in Figure 1.5(a). A binary word stored in a register provides the digital output when conversion is complete. The register drives the D/A converter, the analogue value of which is compared with the analogue input signal and an error generated. If the error is greater than some tolerance value (usually equal to the LSB), it is used by a sequential logic unit to send pulses to the digital register. The conversion then proceeds in a serial manner by reducing the error until it lies within tolerance. The clocked logic circuitry controls the incrementing or decrementing of the register once started and also indicates to the computer when the conversion has been performed.

There are many variants of the closed-loop A/D converter, the differences being primarily concerned with the way the logic unit handles the conversion error and how the digital register is incremented. A commonly

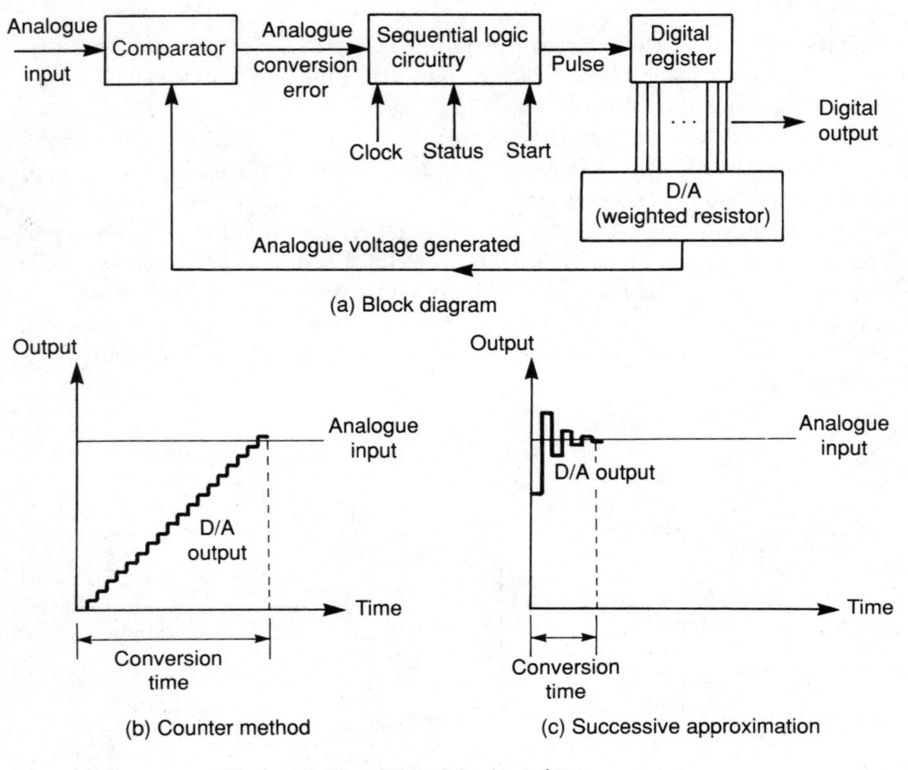

Figure 1.5 Closed-loop A/D converter

used method uses a counter as the register. Here the counter is initialised to zero at the start of each conversion and counts up one pulse at a time, as shown in Figure 1.5(b), until the analogue input and the D/A value are within the error tolerance. Since the register is incremented by one pulse at a time, the overall conversion time can take a considerable number of counts. During this time the analogue input can be held constant by using "hold" devices. Note that the conversion time is not constant and depends upon the analogue input value; the maximum value is, of course, the time required to count 2^n clock pulses, assuming an n-bit conversion, and this can be rather large.

Another technique for closed-loop A/D conversion, shown in Figure 1.5(c), is the successive-approximation method. Here much higher conversion speeds are possible because instead of modifying the LSB of the output register, the most significant bit (MSB) is set first, followed by the next bit, and so on for all the n bits. The conversion time is therefore n clock periods for n-bit conversion, but additional logic circuitry is needed to establish the correct settings for all the bits.

The successive-approximation converter is one of the most widely used methods because of its reliability, ease of design, and relative high speed (see Houpis and Lamont [46]). When compared with the simultaneous A/D converters the closed-loop conversion methods are slower, although 12 bit conversion times of about 5 μs are now possible, which is adequate for most digital control applications.

With several converters of this kind, significant conversion time delays can be introduced into digital computer control system applications. These, together with other sequential processing delays, mean that when analogue signals are to be converted into digital form the conversions can only be performed at discrete instants separated by finite intervals. A typical sampling operation is shown in Figure 1.6, where it can be seen that the digitised signal $y^*(t)$ only appears as an impulsive signal with pulses at the sampling instants $t = kT$ for $k = 0, 1, 2, 3 \ldots$. For most design and analysis purposes it is adequate to assume $y^*(kT) = y(kT)$ for $k = 0, 1, 2, 3 \ldots$, but in practice the A/D conversion involves quantisation rounding errors because of the finite word lengths used. A good discussion on this and other discretisation induced effects can be found in Houpis and Lamont [46].

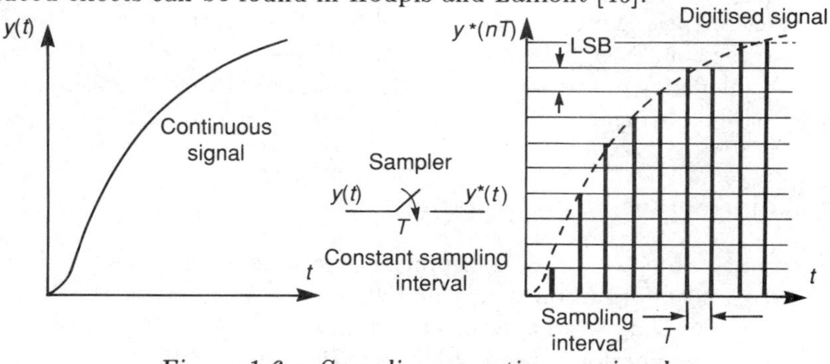

Figure 1.6 Sampling a continuous signal

In computer control applications, impulsive signals are inappropriate for controlling analogue systems since these require an input signal to be present for all time. To overcome this difficulty, hold devices are inserted at the digital-to-analogue interfaces. The simplest device available is a zero-order hold (ZOH), shown in Figure 1.7(a), which holds the output constant at the value fed to it at the last sampling instant; hence a piece-wise constant signal is generated.

Higher order holds are also available (see for example Kuo [70]), which use a number of previous sampling instant values to generate the signal over the current sampling interval. Although requiring extra effort, these approximate the continuous signals more closely and lead to better system performance. For example, a first-order hold, shown in Figure 1.7(b), uses the last two values to linearly extrapolate for some parameter a, where

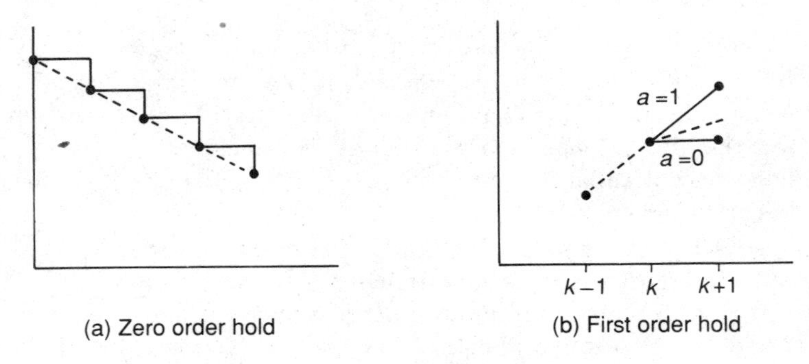

Figure 1.7 Holding devices

$0 < a < 1$, to give the output for the time interval $[kT, (k+1)T]$ as

$$y(kT+t) = y(kT) + a\left\{\frac{y(kT) - y((k-1)T)}{T}\right\}t \qquad \text{for } 0 \le t \le T \quad (1.2)$$

As already mentioned, other interfacing devices may also need to be connected to the computer. These include disk drives, keyboards, display monitors and telemetry modems, and involve special hardware (see for example Bennett [13] , Mellichamp [83]).

Data Representation and Aliasing

The primary objective when sampling a continuous signal, $y(t)$, is that all the relevant information is retained with good accuracy in the sampled signal, $y^*(kT)$. Long word lengths and rapid sampling rates are required in most cases where performance is of utmost importance. However, in many applications the control computer may already be installed and so the only choice available to the engineer will be in the sampling rate selection, since the computer hardware usually dictates the word length (although multiple word-length instructions can be used). The main factors affecting this choice are the bandwidth of the system under study, and the processing power of the computer. The former gives an indication of the slowest rate adequate to prevent loss of information due to a phenomenon called aliasing, where the frequency response of the sampled data system is distorted as higher frequencies are "folded" down to lie within the pass band. To illustrate this, consider a continuous system with bandwidth f_h, as shown in Figure 1.8(a), to which we apply the digitisation procedure.

The highest frequency that needs to be accommodated by the sampling process is f_h . The minimum sampling frequency f_s is then given by the following result:

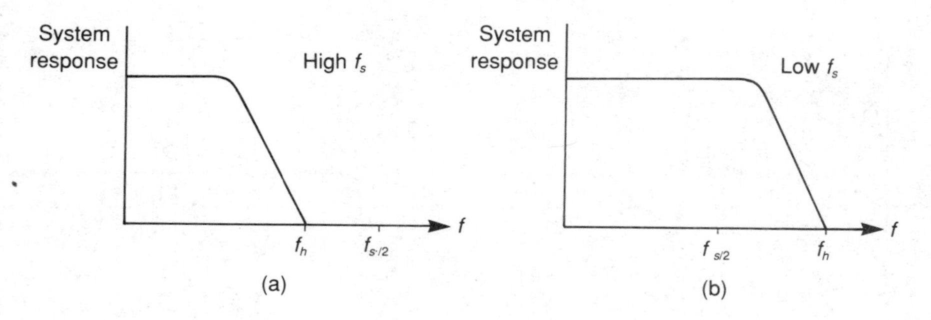

Figure 1.8 *Sampling frequency selection*

Shannon's Sampling Theorem (see Shannon *et al.* [100]; Shannon and Weaver [101])

> Consider a continuous signal $y(t)$ with highest frequency component f_h Hz, and assume that this is sampled at a frequency f_s Hz. Then it is possible to reconstruct $y(t)$ from its sampled version $y^*(kT)$ if, and only if
>
> $$f_s \geq 2f_h \tag{1.3}$$

We may visualise this result if we consider the sampling of a pure sinusoid of frequency f, as shown in Figure 1.9. If the sampling rate is equal to f, that is $T = 1/f$, then the sampler gives a constant output, as shown in Figure 1.9(a). Clearly the sampling rate is inappropriate for the signal being sampled. If T is reduced, as shown in Figure 1.9(b), $y^*(kT)$ appears to be a low frequency sinusoid – but it still fails to represent the original continuous signal accurately, and T must therefore be further reduced. When $T = 1/2f$, shown in Figure 1.9(c), it is straightforward to see that the sampled output has the same frequency as the analogue input, and hence the sampling rate is just adequate for reconstruction purposes as stated in the sampling theorem.

Now, although the frequency of the sinusoid is well reconstructed, the sampler output magnitude is in fact a function of the instants at which the samples are taken, that is, the phase. Clearly this is inadequate for control purposes as such synchronization differences lead to gain variations, and changes in system performance. Hence, for accurate representation, as we shall see later, considerably higher sampling rates than that stated in the Sampling Theorem are normally used (see Figure 1.9(d)).

Returning to our control system of bandwith f_h :

(i) If $f_s \geq 2f_h$, as shown in Figure 1.8(a), no reconstructional problems exist.

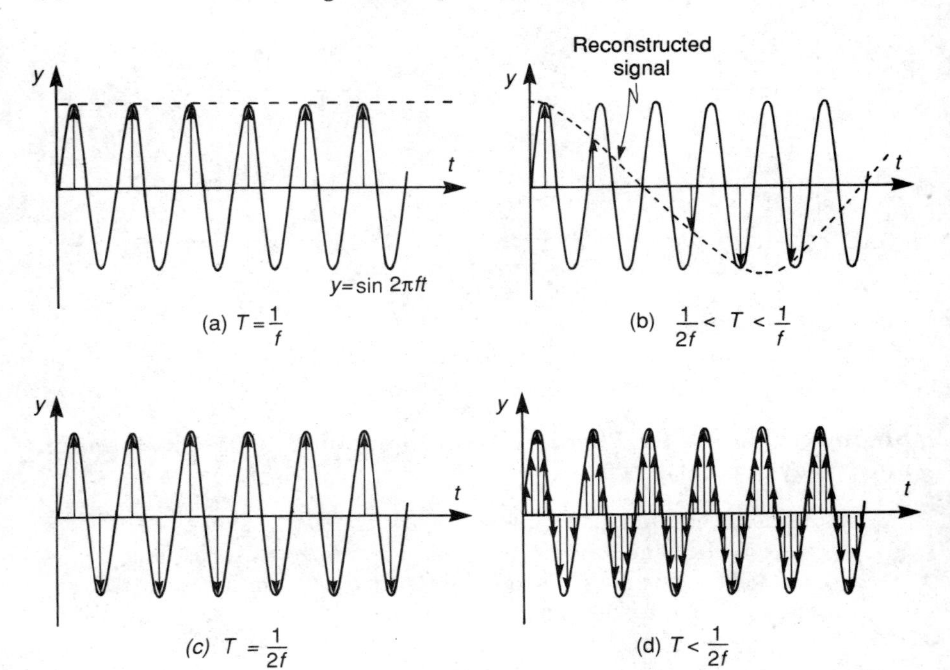

Figure 1.9 Sampling of a sinusoid

(ii) If $f_s < 2f_h$, as shown in Figure 1.8(b), we experience similar problems, as illustrated by the simple sinusoid case, and it is impossible to distinguish between frequencies separated by multiples of the sampling frequency, such as, $f, f + f_s, f + 2f_s$, etc. We say that aliasing has occurred.

The situation can best be illustrated in mathematical terms by considering a frequency, f_ℓ, such that $0 < f_\ell < f_h$, and letting $f = f_\ell + kf_s$ for $k = 0, 1, 2, \ldots$. If this sinusoidal frequency component

$$y(t) = \sin 2\pi ft = \sin 2\pi(f_\ell + kf_s)t$$

is sampled at intervals of T seconds, we have

$$y^*(iT) = \sin 2\pi (f_\ell + kf_s) iT \quad \text{for } i, k = 0, 1, 2, 3, \ldots \tag{1.4}$$

at the sampling instants. Now the sampling frequency $f_s = 1/T$ and so

$$y^*(iT) = \sin 2\pi f_\ell iT \qquad \text{for } i = 0, 1, 2, \ldots \tag{1.5}$$

Therefore sinusoids of frequency $f = f_\ell + kf_s$ for $k = 0, 1, 2, \ldots$ appear as sinusoids of frequency f_ℓ when sampled at f_s Hz, and frequencies higher than $f_s/2$ are mapped down to the interval $[0, f_s/2]$. This folding gives rise to "windows" on the frequency axis, as illustrated in Figure 1.10. When f_s is high (see Figure 1.10(a)) the intervals do not overlap and adequate reconstruction is possible as there are no signals from the higher order windows caused by sampling. However for low sample rates (see Figure 1.10(b)) the frequency windows overlap, and the frequency response of the system is distorted. Since such folding of frequencies can lead to severe distortion, it is common practice to insert low pass, anti-aliasing filters to attenuate the higher frequencies before the signals are sampled.

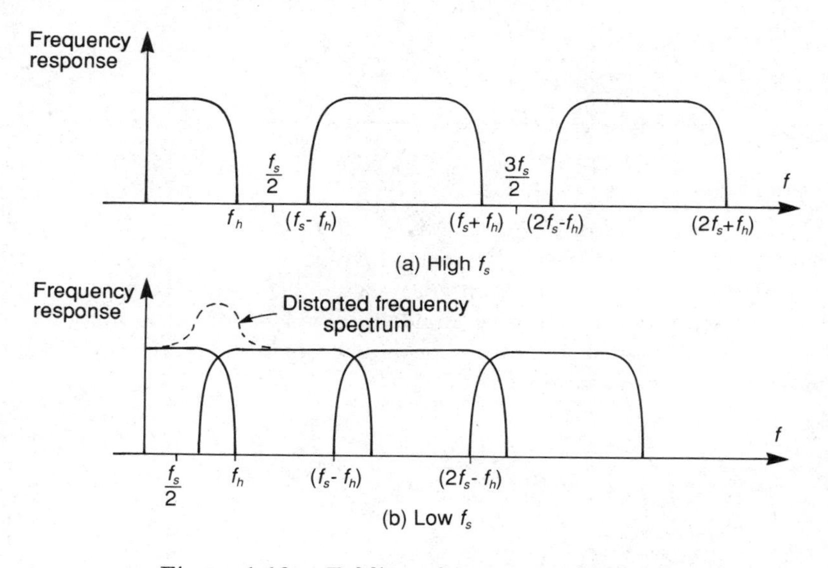

Figure 1.10 Folding of frequency windows

1.4 Summary

It is clear that to achieve good control performance the issues of sample rate selection and the levels of quantisation errors need to be addressed carefully. The sample rate has to be fast enough to avoid aliasing and to give accurate sampled representations of the analogue signals. Similarly the word lengths used in the computer system need to be sufficiently long so that the quantisation errors (LSBs) are as small as possible. In this chapter we have given a general introduction to the area of computer control systems, and have outlined the main interface difficulties to be considered in any application. Some of the points discussed here will be addressed in greater detail in later chapters, but before we close this chapter it is worth mentioning the overall structure of a computer control loop and where the different elements lie. A useful picture of the situation is illustrated in Figure 1.11, where the important elements in a typical control loop are shown.

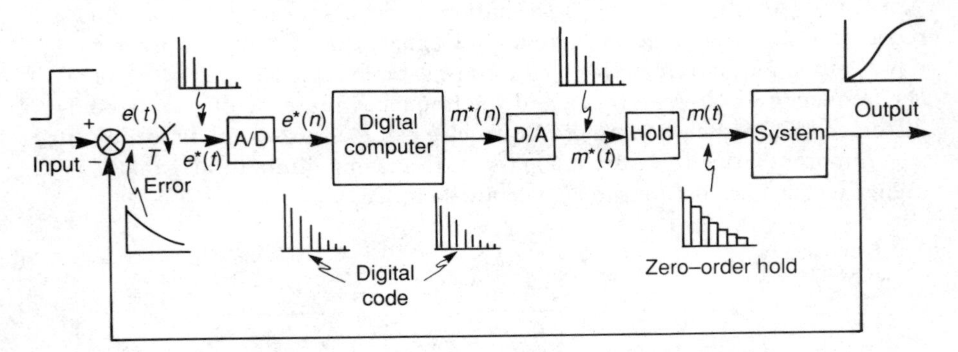

Figure 1.11 Structure of a computer control loop

The continuous error signal $e(t)$ is sampled and converted into digital code $e^*(n)$ ready for processing in the computer. By using a suitable control law, a digital word $m^*(n)$ in then generated and fed to a D/A unit where it is converted into an impulsive analogue signal $m^*(t)$. A holding device is then used to turn $m^*(t)$ into a continuous signal $m(t)$ after which it is applied to the system under control.

2 The z-Transformation

2.1 Introduction

In the previous chapter the concepts of computer control were briefly introduced and some of the considerations in interfacing and discretisation have been discussed. The next important step is to look at design and analysis methods for sampled data systems. We have seen that signals in the control loop are of a discrete nature, and this fact needs to be taken into account in the analysis. For example, consider a continuous function that is sampled as shown in Figure 2.1.

Figure 2.1 Sampling a continuous function

The sampled signal $y^*(t)$ is seen to be a sequence of unit impulses, modulated by the function $y(t)$ at the sampling instants $t = T, 2T, 3T, \ldots$. Using the notation, $\delta(t)$, to represent a unit impulse at time $t = 0$ we have

$$y^*(t) = \sum_{k=0}^{\infty} y(kT)\, \delta(t - kT) \tag{2.1}$$

It is clearly necessary to handle functions of this kind when analysing discrete systems. The Laplace transform (see for example Di Stefano *et al.* [28]) is an invaluable tool for reducing complex differential relations to simple algebraic ones, thereby making analysis easier. Most of the discrete systems analysis, in fact, follows similar lines to the continuous domain analysis, and we therefore start our discussion by posing the following question: "Can the continuous domain Laplace transform techniques be used in studying discrete systems ?" To answer this, let us look at equation (2.1), which, when Laplace-transformed, becomes

$$Y^*(s) = \sum_{k=0}^{\infty} y(kT)\, e^{-ksT} \tag{2.2}$$

since $\mathcal{L}[\delta(t)] = 1$ and $\mathcal{L}[\delta(t-T)] = e^{-sT}$, where $\mathcal{L}[y(t)]$ is the Laplace transform of $y(t)$ (written as $Y(s)$). It turns out that in discrete systems we are faced with the need to handle irrational functions in s of this type, and consequently analysis is rather difficult. However it is possible to introduce a new transformation which reduces these irrational functions to rational ones. This is the so called **z-transform**.

2.2 Definition of the z-Transform

The z-transform is defined by replacing e^{st} terms in expressions of the form (2.2) by z and writing $Y^*(s)$ as $Y(z)$. Hence we define the z-transform as

$$Y(z) = [Y^*(s)]_{z=\exp(sT)} \tag{2.3}$$

and so equation (2.2) becomes

$$Y(z) = \sum_{k=0}^{\infty} y(kT)\, z^{-k} \tag{2.4}$$

For a unit ramp, $y(kT) = kT$ for all $k \geq 0$, and so

$$Y(z) = \sum_{k=0}^{\infty} (kT) z^{-k} = Tz^{-1} + 2Tz^{-2} + 3Tz^{-3} + \cdots \tag{2.5}$$

or

$$Y(z) = \frac{Tz}{(z-1)^2} \tag{2.6}$$

The above two transforms are known, for obvious reasons, as open-form (2.5) and closed-form (2.6). Since $z = e^{sT}$ we see that multiplication by z is in fact equivalent to a pure time advance of T seconds, and multiplication by z^{-1} is equivalent to a pure time delay of T seconds. For example, the function

$$Y(z) = z^{-3} + 5z^{-4} + 12z^{-5} + 15z^{-6} \tag{2.7}$$

has a time response as shown in Figure 2.2. The z-exponent is therefore a time ordering indicator. By using equation (2.4), any continuous time function can be transformed into the z-domain for analysis in the digital environment. The investigations are made easier by converting open-form transforms to closed-form transforms and vice versa as necessary. The open-form, as seen above, gives the time response directly, and the closed-form

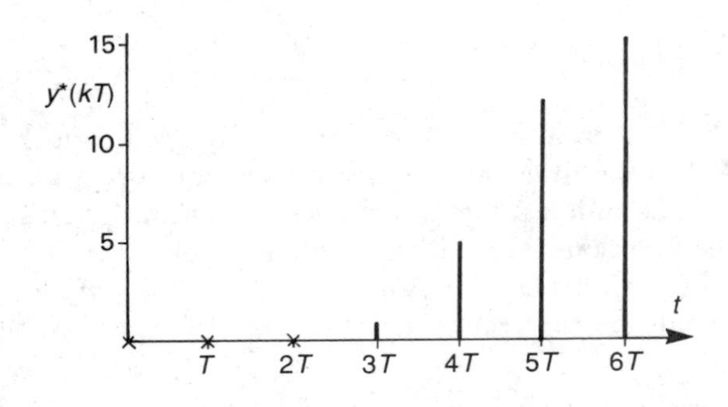

Figure 2.2 *Time response derived from a polynomial in z*

gives a discrete pole-zero transfer function allowing the z-domain studies to proceed. The use of z-transform tables (see appendix A) assists in this analysis.

2.3 z-Transform Properties

It is straightforward to show that the z-transform has the following properties, see for example Kuo [70]; Ogata [89]:

Addition and Subtraction

$$\mathcal{Z}\{y_1(t) \pm y_2(t)\} = Y_1(z) \pm Y_2(z) \tag{2.8}$$

where the expression $\mathcal{Z}\{\cdot\}$ is read as "the z-transform of $\{\cdot\}$".

Multiplication by a Constant

$$\mathcal{Z}\{ay(t)\} = aY(z) \tag{2.9}$$

Translation in Time

$$\mathcal{Z}\{y(t \pm kT)\} = z^{\pm k}Y(z) \tag{2.10}$$

Initial Value Theorem

$$\lim_{t \to 0} y(t) = \lim_{z \to \infty} Y(z) \tag{2.11}$$

Final Value Theorem

$$\lim_{t \to \infty} y(t) = \lim_{z \to 1} \left(\frac{z-1}{z}\right) Y(z) \tag{2.12}$$

These properties simplify and enable the z-transformation method to be applied to the study of discrete system behaviour.

2.4 Inverse z-Transform

As in the continuous case, it is often necessary to transform z-domain expressions back into the time domain, so that system responses are available for assessment. The simplest method of performing this inverse z-transform is to expand the function into an infinite series in powers of z^{-1} and thereby obtain its open-form z-transform. Then as shown in Figure 2.2, the coefficients of the expansion represent the values of the function at the sampling instants.

Another commonly used method is to use z-transform tables for inverse z-transforming standard functions (see appendix A). Complicated expressions can be simplified, using partial fractions, into standard terms before consulting the z-transform tables. It is worth noting that the partial fraction expansions, although similar to the Laplace transformation case, do differ in one minor respect. In the continuous case a function $Y(s)$, where

$$Y(s) = \frac{a_{n-1}s^{n-1} + a_{n-2}s^{n-2} + \cdots + a_1 s + a_0}{(s+p_1)(s+p_2)\cdots(s+p_n)} \tag{2.13}$$

is inverse Laplace-transformed by expanding it as

$$Y(s) = \frac{A_1}{s+p_1} + \frac{A_2}{s+p_2} + \cdots + \frac{A_n}{s+p_n} \tag{2.14}$$

where $p_1, p_2, \cdots, p_n$ are the poles of $Y(s)$ and $A_1, A_2, \cdots, A_n$ are the residues of $Y(s)$ at these poles. Assuming all the p_i values are positive, the inverse Laplace-transform of $Y(s)$ is then

$$y(t) = A_1\exp(-p_1 t) + A_2\exp(-p_2 t) + \cdots + A_n\exp(-p_n t) \tag{2.15}$$

In the discrete case, $Y(z)$ is not expanded into a form similar to equation (2.14). The reason is that the inverse z-transform of a term such as $A/(z+a)$ is not found in the z-transform table, although it represents a delayed pulse train with exponentially decaying amplitude if a is positive. Furthermore, it is known that the inverse transform of $Az/\left(z-e^{-aT}\right)$ is Ae^{-akT}. Therefore it is more appropriate to expand the function $F(z)/z$ using partial fractions. After the expansion, both sides of $F(z)/z$ are multiplied by z to obtain $F(z)$.

For functions with no zeros at $z = 0$, the corresponding time series has time delays and the partial fraction expansion of $F(z)$ is performed in the usual way, that is,

$$Y(z) = \frac{A_1}{z+p_1} + \frac{A_2}{z+p_2} + \cdots + \frac{A_n}{z+p_n} \tag{2.16}$$

We then let

$$Y_1(z) = zY(z) = \frac{A_1 z}{z + p_1} + \frac{A_2 z}{z + p_2} + \cdots + \frac{A_n z}{z + p_n} \tag{2.17}$$

Once the inverse z-transform of $Y_1(z)$, $y_1(kT)$ is determined, it is straightforward to see that the inverse z-transform of $Y(z)$ is related to $y_1(kT)$ through

$$y(kT) = y_1((k-1)T) \tag{2.18}$$

Another alternative approach is obtained by using the following inverse transform formula, which is derived from the definitions of the Laplace and z-transforms (see Kuo [70])

$$y(kT) = \frac{1}{2\pi j} \int_{\Gamma} Y(z) z^{k-1} \mathrm{d}z \tag{2.19}$$

where Γ is a closed path (usually a circle) in the z-plane which encloses all the singularities of $Y(z) z^{k-1}$. The integrand in equation (2.19) can be evaluated using the residue theorem of complex variable theory (see Kuo [70]), to give

$$y(kT) = \sum \left\{ \text{residue of } Y(z) z^{k-1} \text{ at the poles of } Y(z) \right\} \tag{2.20}$$

Example

Consider $Y(z)$ where

$$Y(z) = \frac{0.3z}{(z-1)(z-0.7)} \tag{2.21}$$

We will inverse z-transform this using the three methods mentioned above.

Power-series expansion

Using long division $Y(z)$ can be expanded as a series in z^{-1} to give

$$
\begin{array}{r|l}
 & 0.3z^{-1} + 0.51z^{-2} + 0.657z^{-3} + 0.760z^{-4} + 0.832z^{-5} + \cdots \\
\hline
z^2 - 1.7z + 0.7 & 0.3z \\
 & 0.3z - 0.51 + 0.210z^{-1} \\
\cline{2-2}
 & \qquad 0.51 - 0.210z^{-1} \\
 & \qquad 0.51 - 0.867z^{-1} + 0.357z^{-2} \\
\cline{2-2}
 & \qquad\qquad 0.657z^{-1} - 0.357z^{-2} \\
 & \qquad\qquad 0.657z^{-1} - 1.117z^{-2} + 0.460z^{-3} \\
\cline{2-2}
 & \qquad\qquad\qquad 0.760z^{-2} - 0.460z^{-3} \\
 & \qquad\qquad\qquad 0.760z^{-2} + 1.292z^{-3} + 0.532z^{-4} \\
\cline{2-2}
 & \qquad\qquad\qquad\qquad 0.832z^{-3} - 0.532z^{-4} \\
 & \qquad\qquad\qquad\qquad \vdots \qquad\qquad \vdots
\end{array}
$$

Therefore we have

$$Y(z) = 0.3z^{-1} + 0.51z^{-2} + 0.657z^{-3} + 0.760z^{-4} + \cdots \tag{2.22}$$

which, when written in the time domain, is

$$y(kT) = 0.3\delta(t - T) + 0.51\delta(t - 2T) + 0.657\delta(t - 3T) + \cdots \tag{2.23}$$

where $\delta(t - kT)$ represents a unit impulse at $t = kT$.

Partial fraction expansion

As discussed above we can use partial fractions to expand $Y(z)$ in equation (2.21) to give

$$Y(z) = \frac{z}{z-1} - \frac{z}{z-0.7} \tag{2.24}$$

From tables we can inverse z-transform term by term to yield

$$y(kT) = 1 - 0.7^k \qquad \text{for} \quad k = 0, 1, 2, 3, \ldots \tag{2.25}$$

and so $y(kT)$ is an exponentially increasing signal with a steady state value of unity. It is straightforward to deduce that

$$
\begin{array}{lcllcl}
y(0) & = & 0 & y(6T) & = & 0.8824 \\
y(T) & = & 0.3 & y(7T) & = & 0.9176 \\
y(2T) & = & 0.51 & y(8T) & = & 0.9424 \\
y(3T) & = & 0.657 & y(9T) & = & 0.9596 \\
y(4T) & = & 0.7599 & y(10T) & = & 0.9718 \\
y(5T) & = & 0.83193 & y(11T) & = & 0.9802
\end{array} \tag{2.26}
$$

This agrees with the result obtained above.

Inverse transform formula

As mentioned above, the integrand in the inverse transform formula can be evaluated using the residue theorem, and so

$$\begin{aligned} y(kT) &= \sum \left\{ \text{res of } \frac{0.3z}{(z-1)(z-0.7)} z^{k-1} \text{ at } z = 1 \text{ and } 0.7 \right\} \\ &= \left[\frac{0.3z^k}{z-0.7} \right]_{z=1} + \left[\frac{0.3z^k}{z-1} \right]_{z=0.7} \\ &= 1 - 0.7^k \quad \text{for } k = 0, 1, 2, 3, \ldots \end{aligned} \tag{2.27}$$

which again agrees with the result obtained previously.

A point to remember when performing the inverse z-transform is that the function values are defined only at the sampling instants. The transform function does not contain any information for the time intervals between samples, and so the sample rate must be chosen such that it is adequate for the signal being represented. An example of two quite different time functions, having the same z-transform, is illustrated in Figure 2.3. Hence,

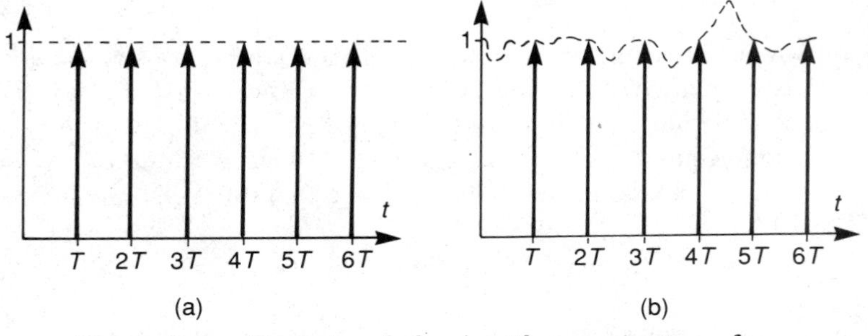

Figure 2.3 *Two signals having the same z-transform*

starting with $Y(z)$ it may be difficult to deduce the time domain properties uniquely. The modified z-transform, by permitting the calculation of $y(t)$ in the intervals $kT \leq t \leq (k+1)T$, for $k = 0, 1, 2, \ldots$, alleviates this limitation as discussed in section 2.7.

2.5 Pulse or z-Transfer Functions of Systems

The z-transfer functions for discrete system analysis can be arrived at by starting with the continuous system illustrated in Figure 2.4(a). Introducing an ideal sampler at the input side, we see that the output is defined by

$$Y(s) = G(s)\, U^*(s) \tag{2.28}$$

Since z-domain analysis can only define the output at the sampling instants, we assume that a fictitious synchronised sampler exists and thereby obtain

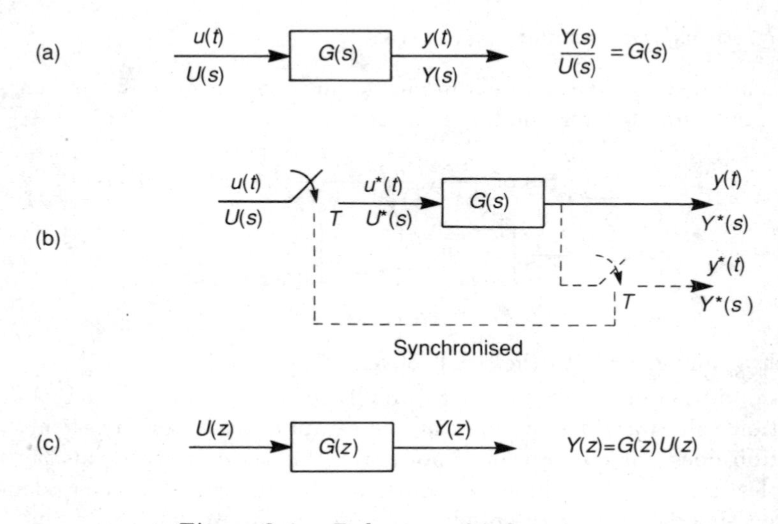

Figure 2.4 Pulse transfer functions

the sampled output transform. Here $Y(s)$ is the product of two transforms, where one is periodic, with period $2\pi/T$ ($U^*(s)$), and the other is not ($G(s)$). It can be shown that if such a product transform is sampled, the periodic transform comes out as a factor of the result. This is the most important relation for the block diagram analysis of sampled-data systems, namely

$$Y^*(s) = (G(s)\,U^*(s))^* = G^*(s)\,U^*(s) \tag{2.29}$$

We will prove (2.29) in the time domain and start by inverse Laplace-transforming $Y(s)$ to give

$$y(t) = \mathcal{L}^{-1}\{G(s)\,U^*(s)\} = \int_0^t g(t-\tau)\,u^*(\tau)\,\mathrm{d}\tau \tag{2.30}$$

This can be further simplified by using equation (2.1) on u^*, so that

$$\begin{aligned} y(t) &= \int_0^t g(t-\tau)\sum_{k=0}^{\infty} u(kT)\,\delta(\tau-kT)\,\mathrm{d}\tau \\ &= \sum_{k=0}^{\infty}\int_0^t g(t-\tau)\,u(kT)\,\delta(\tau-kT)\,\mathrm{d}\tau \\ &= \sum_{k=0}^{\infty} g(t-kT)\,u(kT) \end{aligned} \tag{2.31}$$

Then using equation (2.4), the z-transform of $y(t)$ becomes

$$\begin{aligned} Y(z) &= \sum_{n=0}^{\infty} y(nT)\, z^{-n} = \sum_{n=0}^{\infty} \left\{ \sum_{k=0}^{\infty} g(nT - kT)\, u(kT) \right\} z^{-n} \\ &= \sum_{m=0}^{\infty} \sum_{k=0}^{\infty} g(mT)\, u(kT)\, z^{-(m+k)} \end{aligned} \tag{2.32}$$

where $m = n - k$. Therefore

$$\begin{aligned} Y(z) &= \sum_{m=0}^{\infty} g(mT)\, z^{-m} \sum_{k=0}^{\infty} u(kT)\, z^{-k} \\ &= G(z)\, U(z) \end{aligned} \tag{2.33}$$

is the z-transform of the output. Since the z-transform is the starred Laplace transform with e^{sT} replaced by z, we may express (2.33) as

$$Y^*(s) = G^*(s)\, U^*(s) \tag{2.34}$$

which is the result required. We have thus shown that by taking the starred Laplace transform of both sides of equation (2.28) we obtain equation (2.34), which has (2.33) as its z-transform.

Considering equation (2.33), when the input $u(t)$ is a unit impulse at time $t = 0$, we have, using equation (2.4), that

$$U(z) = \sum_{k=0}^{\infty} u(kT)\, z^{-k} = \sum_{k=0}^{\infty} \delta(0)\, z^{-k} \tag{2.35}$$

and so $U(z) = 1$ for a unit impulse. In this instance $Y(z) = G(z)$, and so $G(z)$ is commonly referred to as the pulse transfer function. Note that the output can still be a continuous function, but the standard z-domain analysis only defines it at the sampling instants.

We will now use an example to illustrate how the pulse or z-transfer function of a continuous system is obtained. Consider

$$\frac{Y(s)}{U(s)} = G(s) = \frac{1}{s(s+1)} \tag{2.36}$$

Putting $G(s)$ into partial fractions gives

$$G(s) = \frac{1}{s} - \frac{1}{s+1} \tag{2.37}$$

and using z-transform tables for each term gives the z-transfer function as

$$G(z) = \frac{z}{z-1} - \frac{z}{z-e^{-T}} = \frac{z\left(1-e^{-T}\right)}{(z-1)\left(z-e^{-T}\right)} \tag{2.38}$$

where T is the sampling interval. Hence the digitised system has the transfer function

$$G(z) = \frac{Y(z)}{U(z)} = \frac{z\left(1 - e^{-T}\right)}{(z-1)\left(z - e^{-T}\right)} \tag{2.39}$$

2.6 Discrete Approximations

Whenever we introduce sampling into continuous operations, the continuous transfer functions need to be transformed into z-domain transfer functions. The conversion can be carried out by substituting

$$z = e^{sT} \quad \text{or} \quad s = (1/T)\ln z \tag{2.40}$$

to give the "exact" z-transform. However this substitution is awkward and leads to complicated terms which are difficult to handle. It turns out that far simpler "approximate" transformations can be used to give adequate description for most cases. The two main methods in current use are

(i) numerical integration, and
(ii) pole-zero mapping,

which we now briefly describe.

Numerical Integration

In this technique, differential equations are transformed into difference equations by using some numerical method to approximate the integration operation. To illustrate the technique, consider

$$\frac{Y(s)}{U(s)} = G(s) = \frac{a}{s+a} \tag{2.41}$$

Cross multiplying gives

$$sY(s) + aY(s) = aU(s) \tag{2.42}$$

Inverse Laplace-transforming, term by term, and assuming zero initial conditions, results in

$$\frac{\mathrm{d}y(t)}{\mathrm{d}t} + ay(t) = au(t) \tag{2.43}$$

For a digital representation, such a differential equation needs to be solved at the sampling instants, and so we need to calculate

$$y(kT) = y((k-1)T) + \int_{(k-1)T}^{kT} (-ay(t) + au(t))\,\mathrm{d}t \qquad \text{for } k = 1, 2, 3, \ldots \tag{2.44}$$

The integrand on the right-hand side can be solved numerically to give a digital approximation. Several methods for doing this are available but we shall only present some of the simpler ones here.

Forward Rectangular Rule

Consider the function $f(t)$ defined by $f(t) = -ay(t) + au(t)$. In the forward rectangular rule approach the area under the $f(t)$ curve is approximated by assuming that it remains constant at its value at the left-hand end of the time interval, as shown in Figure 2.5(a), so that

$$y(kT) = y((k-1)T) + T\{-ay((k-1)T) + au((k-1)T)\} \quad \text{for } k = 1, 2, 3, \ldots \tag{2.45}$$

As multiplication by z^{-1} represents a delay of T seconds, we have

$$y(kT) = (1 - aT)\, z^{-1} y(kT) + aTz^{-1} u(kT) \tag{2.46}$$

Since $y(kT)$, and $u(kT)$ only appear at the discrete sampling instants when $k = 0, 1, 2, \ldots$, we can replace them by their z-transforms. Therefore we have

$$\frac{Y(z)}{U(z)} = \frac{a}{\frac{z-1}{T} + a} \tag{2.47}$$

Comparing equation (2.41) with (2.47) we can see that the forward rectangular rule gives a digital approximation by setting

$$s = \frac{z-1}{T} \tag{2.48}$$

Backward Rectangular Rule

The backward rule assumes the function value is constant at its value at the right-hand end of the integration interval, as shown in Figure 2.5(b).

Hence, using this method

$$y(kT) = y((k-1)T) + Tf(kT) \tag{2.49}$$

which, using a similar approach to that above, leads to

$$\frac{Y(z)}{U(z)} = \frac{a}{\frac{z-1}{Tz} + a} \tag{2.50}$$

Therefore the backward approximation gives a discrete transfer function by setting

$$s = \frac{z-1}{Tz} \tag{2.51}$$

Trapezium Rule

It is clear that one of the above methods over-estimates the integrand while the other under-estimates it. A closer approximation is obtained by taking

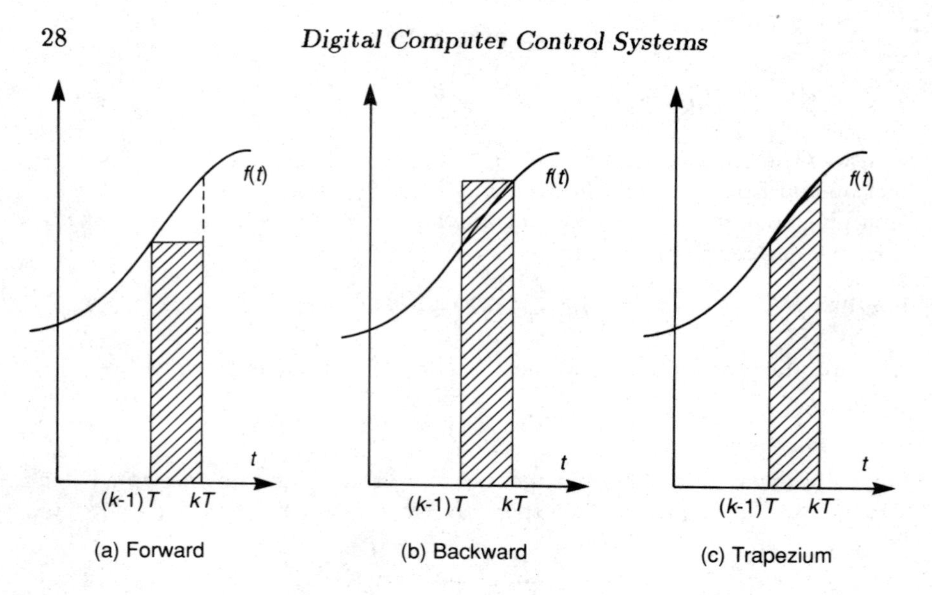

Figure 2.5 Numerical integration methods

the average of the two end values, that is, the area of the trapezium under the f curve, as shown in Figure 2.5(c), so that

$$y(kT) = y((k-1)T) + \frac{T}{2}\{f(kT) + f((k-1)T)\} \tag{2.52}$$

and hence

$$\frac{Y(z)}{U(z)} = \frac{a}{\frac{2}{T}\frac{z-1}{z+1} + a} \tag{2.53}$$

Hence the third and best approximation of the three is obtained by setting

$$s = \frac{2}{T}\frac{z-1}{z+1} \tag{2.54}$$

In digital control this method is known as Tustin's rule, in recognition of his work in the area. The transformation (2.54) is also called the bilinear transformation from consideration of its mathematical form.

We illustrate the use of the three approximations by the aid of an example. Consider

$$\frac{Y(s)}{U(s)} = \frac{1}{s^2 + 0.4s + 0.68} \tag{2.55}$$

which can be factorised to give

$$\frac{Y(s)}{U(s)} = \frac{1}{(s + 0.2 + j0.8)(s + 0.2 - j0.8)} \tag{2.56}$$

The continuous system has its poles at $s = -0.2 \pm j0.8$, and so has a natural undamped frequency, $\omega_n \approx 0.8$ rad/s $= 0.13$ Hz. Sampling with $T = 1$ s should be adequate. Using the different numerical methods gives rise to the following discrete approximations:

(i) forward rule gives discrete poles at $z = 0.8 \pm j0.8$ (an unstable approximation – see section 3.3 in chapter 3);

(ii) backward rule gives poles at $z = 0.58 \pm j0.38$;

(iii) Tustin's rule gives poles at $z = 0.61 \pm j0.58$.

The "exact" transformation $s = (1/T) \ln(z)$ gives poles at $z = 0.57 \pm j0.59$.

Clearly the three methods give varying approximations, and it is advisable to ensure that accuracy and stability are retained when transforming from $G(s)$ to $G(z)$ in this manner. Further discussion on why these variations occur is presented in chapter 3.

Pole-Zero Mapping

As the name implies, this method gives a digital approximation by transforming the poles and zeros in the s-plane directly into poles and zeros in the z-domain using the definition of z. The gain of the discrete transfer function, $G(z)$, is then modified to match that of the initial continuous transfer function, $G(s)$, at a suitable frequency in the passband of the system. In most control applications these gains are matched at zero frequency, but obviously for bandpass systems the frequency selected needs to be in the bandwidth. The pole-zero mapping approach is based on the following set of rules:

(i) map all poles according to $z = e^{sT}$, so if $G(s)$ has a pole at $s = -a$ then $G(z)$ has a pole at $z = e^{-aT}$;

(ii) map all zeros according to $z = e^{sT}$;

(iii) map all zeros of $G(s)$ at $s = \infty$ to $z = -1$ in $G(z)$, i.e. add $(z+1)$, $(z+1)^2$, etc. terms to the numerator of $G(z)$ so that the orders of the numerator and denominator are equal;

(iv) match the gains of $G(s)$ and $G(z)$ at a suitable frequency. For example at zero frequency, $s = 0$, and $z = 1$, and so

$$[\ G(s)\ ]_{s=0} = [\ G(z)\ ]_{z=1} \tag{2.57}$$

The reason for rule (iii) becomes obvious if, in the s-plane, we look at frequencies increasing from $j\omega = 0$. In the z-plane, these are mapped onto the unit circle starting at $z = e^{j0} = 1$, moving anticlockwise until $z = e^{j\pi} = -1$. Frequencies between $\omega = \pi$ and 2π map onto the negative half of the unit circle and higher frequencies repeatedly generate this unit circle. Thus the point $z = -1$ represents, in a real way, the highest frequency possible in the discrete transfer function, and so it is appropriate that if $G(s)$ is zero at the highest (continuous) frequency, $G(z)$ should also have this property. Hence zeros at $s = \infty$ are mapped to $z = -1$.

If a unit delay in the digital unit pulse response is desirable for any reason, such as when computation time is needed to process each sample, a modified pole-zero mapping is available. Here one zero of $G(s)$ at $s = \infty$ is mapped into $z = \infty$ and so the order of the numerator of $G(z)$ is one less than its denominator order. In this instance the series expansion of $G(z)$ in powers of z^{-1} will have no constant term, and hence the $g(kT)$ has a delay of one sampling interval to a unit impulse.

Example

Use the pole-zero mapping method to digitise the system whose continuous transfer function is

$$D(s) = \frac{50}{(s+5)(s+10)} \tag{2.58}$$

with $T = 1/15$ s.

Application of the above procedure gives the following:

- $G(z)$ has poles at $z = e^{-5T}$ and e^{-10T}, that is at $z = 0.715$ and 0.513
- $G(s)$ has a 2 zeros at $s = \infty$ and so $G(z)$ has two zeros at $z = -1$.

Hence

$$G(z) = K\frac{(z+1)^2}{(z-0.715)(z-0.513)} \tag{2.59}$$

Matching low frequency (DC) gains we have

$$\left[\frac{50}{(s+5)(s+10)}\right]_{s=0} = K\left[\frac{(z+1)^2}{(z-0.715)(z-0.513)}\right]_{z=1} \tag{2.60}$$

and so $K = 0.0347$, giving

$$G(z) = 0.0347\frac{(z+1)^2}{(z-0.715)(z-0.513)} \tag{2.61}$$

Starting with a continuous system, we have presented several methods for obtaining discrete equivalents. These included the numerical integration methods where the continuous transfer function, $G(s)$, is represented as a differential equation and is used to derive a difference equation by approximating the integration. When $G(s)$ is not in factored form the integration method is straightforward to apply, whereas the pole-zero mapping technique is suited to factorised transfer functions. All the digitisation methods give varying degrees of quantisation errors in different applications, and so an individual method cannot be said to be best in all cases (see Ogata [88]; Franklin and Powell [35]). For most cases however, since the pole-zero mapping method uses simple algebra, it tends to be the preferred approach.

2.7 Modified z-Transform

As mentioned earlier, the z-transform method allows us to determine the responses only at the sampling instants. In most cases this is not a serious limitation if the sampling theorem is satisfied, since the output will not vary much between any two consecutive sampling instants. However, in cases when signals cannot be represented adequately by their values at the sampling instants, we must acquire knowledge on the system responses between sampling instants. It is possible to obtain these values by using the modified z-transform, where fictitious delays are inserted at the output of the system, as well as the fictitious output sampler. By varying the amount of delay, we can obtain the output at any time between two consecutive sampling instants. The modified z-transform is not only useful for determining intermediate values in this way, but it can also be used in analysing sampled data systems containing pure dead time delays or transportation lags.

Consider the system shown in Figure 2.6, where a fictitious delay of $(1-m)T$ seconds, $0 \leq m \leq 1$ is inserted at the output. By varying m between 0 and 1, the output

$$y(t) \text{ at } t = kT \to (k-1)T \text{ for } k = 0, 1, 2, 3, \ldots \tag{2.62}$$

can be obtained.

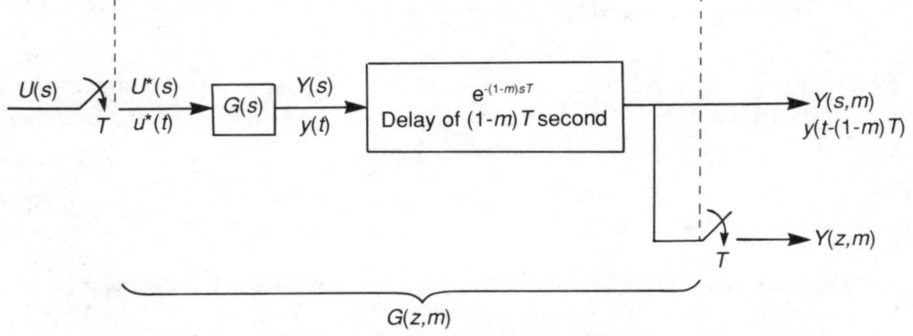

Figure 2.6 Modified z-transform using fictitious time delays

The sampled output of the fictitious time delay is given by

$$y^*(t-(1-m)T) = \sum_{k=0}^{\infty} y(kT-(1-m)T)\,\delta(t-kT) \quad k = 0, 1, 2, 3, \ldots \tag{2.63}$$

The z-transform of $y^*(t-(1-m)T)$ can be expressed as

$$\mathcal{Z}\{y^*(t-(1-m)T)\} = Y(z,m) \tag{2.64}$$

where $Y(z,m)$ is defined as the modified z-transform of $y(t)$, and is given by

$$Y(z,m) = \left[\mathcal{L}\left\{y(t-(1-m)T)\right\} * \mathcal{L}\left\{\sum_{k=0}^{\infty}\delta(t-kT)\right\}\right]_{z=\exp(sT)} \tag{2.65}$$

where $*$ represents the normal Laplace convolution. Hence

$$Y(z,m) = \left[Y(s)e^{-(1-m)sT} * \frac{1}{1-e^{-sT}}\right]_{z=\exp(sT)} \tag{2.66}$$

The convolution in equation (2.66) is written as

$$Y(z,m) = \frac{1}{2\pi j}\left[\int_{c-j\infty}^{c+j\infty} Y(p)e^{-(1-m)pT}\frac{1}{1-e^{-(s-p)T}}\mathrm{d}p\right]_{z=\exp(sT)} \tag{2.67}$$

The line integration in equation (2.67) can be evaluated along the line from $c-j\infty$ to $c+j\infty$ and the semi-circle of infinite radius in either the left half or the right half of the complex p-plane. Considering integration along the left half plane and using the residue theorem of complex-variable analysis (see Kuo [70]; Ogata [89] for details), equation (2.67) is written

$$Y(z,m) = z^{-1}\sum\left\{\text{residue of } \frac{Y(s)\,e^{msT}\,z}{z-e^{-sT}} \text{ at the poles of } Y(s)\right\} \tag{2.68}$$

which gives a method for determining the modified z-transform for any continuous $Y(s)$. Note that the modified z-transform, $Y(z,m)$, and the ordinary z-transform, $Y(z)$, are related by

$$Y(z) = \lim_{m\to 0} zY(z,m) \tag{2.69}$$

As in the case of the z-transform, the modified z-transform $Y(z,m)$ can be expanded as an infinite series in z^{-1}, to give

$$Y(z,m) = y_0(m)z^{-1} + y_1(m)z^{-2} + y_2(m)z^{-3} + \cdots \tag{2.70}$$

Multiplying both sides of equation (2.70) by z gives

$$zY(z,m) = y_0(m) + y_1(m)z^{-1} + y_2(m)z^{-2} + \cdots \tag{2.71}$$

where $y_k(m)$ represents the value of $y(t)$ between $t = kT$ and $t = (k+1)T$, for $k = 0, 1, 2, 3, \ldots$, or

$$y_k(m) = y((k+m)T) \tag{2.72}$$

Note that if $y(t)$ is continuous, then

$$\lim_{m \to 1} y_{k-1}(m) = \lim_{m \to 0} y_k(m) \tag{2.73}$$

The left-hand side of equation (2.73) gives the values $y(0_-), y(T_-), \ldots$, and the right-hand side gives the values $y(0^+), y(T^+), y(2T^+), \ldots$. If the output is continuous, then $y(kT_-) = y(kT^+)$.

We illustrate the modified z-transform technique for obtaining inter-sample values by considering a simple closed-loop sampled-data system example shown in Figure 2.7. Let us initially determine the step response of

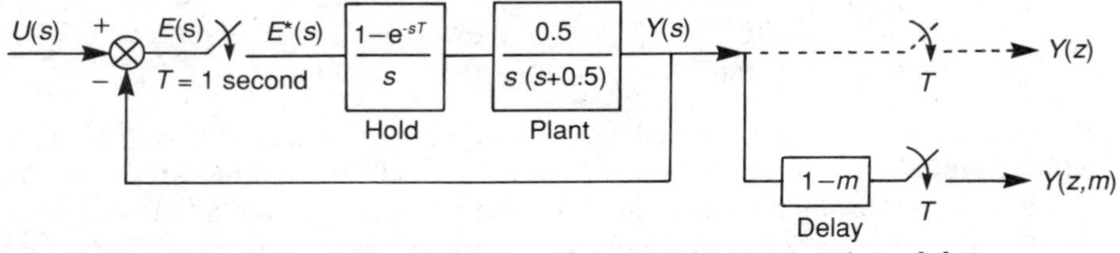

Figure 2.7 Closed-loop system with fictitious time delay

this system using the standard z-transform method. The following equations can be written directly from the block diagram:

$$Y(s) = G(s)E^*(s) \tag{2.74}$$

$$E(s) = U(s) - G(s)E^*(s) \tag{2.75}$$

where $G(s)$ is the combined transfer function of the plant and zero-order hold unit. Sampling these and combining gives

$$Y^*(s) = \frac{G^*(s)}{1 + G^*(s)} U^*(s) \tag{2.76}$$

which written in z-transform terms equals

$$Y(z) = \frac{G(z)}{1 + G(z)} U(z) \tag{2.77}$$

Now

$$G(s) = \frac{\left(1 - e^{-sT}\right) 0.5}{s^2 (s + 0.5)} \tag{2.78}$$

and so, from tables we can deduce that

$$G(z) = \frac{0.213z + 0.18}{z^2 - 1.607z + 0.607} \tag{2.79}$$

giving the digital output as

$$Y(z) = \frac{0.213z + 0.18}{z^2 - 1.394z + 0.787} U(z) \tag{2.80}$$

When the input is a unit step, $U(z) = z/(z-1)$, and the output response values, at the sampling instants, can be obtained via long division as

$$\begin{array}{lll} y(0) = 0 & y(10) = 0.702 \\ y(1) = 0.213 & y(11) = 0.831 \\ y(2) = 0.69 & y(12) = 0.998 \\ y(3) = 1.188 & y(13) = 1.131 \\ y(4) = 1.505 & y(14) = 1.184 \\ y(5) = 1.556 & y(15) = 1.153 \\ y(6) = 1.378 & y(16) = 1.069 \\ y(7) = 1.088 & y(17) = 0.975 \\ y(8) = 0.826 & y(18) = 0.911 \\ y(9) = 0.688 & y(19) = 0.896 \end{array} \tag{2.81}$$

Having calculated the standard discrete response, assume this is not adequate and that inter-sampling values of the output are required. Such values can be obtained by positioning a fictitious time delay at the output, as shown in Figure 2.7, and using the modified z-transform as discussed above. Since

$$Y(s) = G(s)\,E^*(s) \tag{2.82}$$

the modified z-transform of the output equals

$$Y(z,m) = G(z,m)\,E(z) \tag{2.83}$$

The last relation holds because the modified z-transform of a transfer function already in sampled form is the z-transform of the function (see for example Kuo [70]). Hence it is straightforward to see that for our closed-loop system we have

$$Y(z,m) = \frac{G(z,m)}{1+G(z)}U(z) \tag{2.84}$$

The modified z-transform of $G(s)$ is obtained from equation (2.68) as follows:

$$\begin{aligned} G(z,m) &= z^{-1}\sum\left\{\text{residue of } \frac{G(s)\,e^{ms}\,z}{z-e^{-s}} \text{ at the poles of } G(s)\right\} \\ &= \frac{z-1}{z^2}\left\{\left[\text{res of } \frac{0.5}{s^2(s+0.5)}\frac{e^{ms}\,z}{(z-e^{-s})} \text{ at double pole } s=0\right]\right. \\ &\quad \left. + \left[\text{res of } \frac{0.5}{s^2(s+0.5)}\frac{e^{ms}\,z}{(z-e^{-s})} \text{ at pole } s=-0.5\right]\right\} \\ &= \frac{z-1}{z^2}\left\{\frac{1}{(2-1)!}\lim_{s\to 0}\frac{\mathrm{d}}{\mathrm{d}s}\left[s^2\frac{0.5}{s^2(s+0.5)}\frac{e^{ms}z}{(z-e^{s})}\right]\right. \\ &\quad \left. + \lim_{s\to -0.5}(s+0.5)\frac{0.5}{s^2(s+0.5)}\frac{e^{ms}z}{(z-e^{s})}\right\} \end{aligned}$$

$$G(z,m) = \frac{z-1}{z^2}\left[\frac{mz^2 - mz - 2z^2 + 3z}{(z-1)^2} + \frac{2e^{-0.5m}z}{z - 0.607}\right]$$

This gives

$$G(z,m) = \frac{z^2 a_2 + z a_1 + a_0}{z(z-1)(z-0.607)} \tag{2.85}$$

where

$$\begin{aligned} a_0 &= 0.607m - 1.821 + 2e^{-0.5m} \\ a_1 &= 4.214 - 1.607m - 4e^{-0.5m} \\ a_2 &= m - 2 + 2e^{-0.5m} \end{aligned} \tag{2.86}$$

Referring to equation (2.84) and noting that $U(z) = z/(z-1)$, we have

$$\begin{aligned} Y(z,m) &= \frac{G(z,m)}{1+G(z)}\frac{z}{z-1} \\ &= \frac{z^2 a_2 + z a_1 + a_0}{z^3 - 2.394z^2 + 2.181z - 0.787} \end{aligned} \tag{2.87}$$

Hence, for $m = 0.5$, we have

$$Y(z,0.5) = \frac{0.058z^2 + 0.295z + 0.04}{z^3 - 2.394z^2 + 2.181z - 0.787} \tag{2.88}$$

Expanding equation (2.88) as an infinite series in z^{-1} gives

$$Y(z,0.5) = y_0(0.5)z^{-1} + y_1(0.5)z^{-2} + y_2(0.5)z^{-3} + \cdots \tag{2.89}$$

where $y_k(0.5) = y((k+0.5)T) = y(k+0.5)$ for $k = 0, 1, 2, \ldots$. Using long division the values of $y_k(0.5)$ are found to be

$$\begin{aligned} y(0.5) &= 0 & y(10.5) &= 0.6763 \\ y(1.5) &= 0.0576 & y(11.5) &= 0.7552 \\ y(2.5) &= 0.4332 & y(12.5) &= 0.9135 \\ y(3.5) &= 0.9515 & y(13.5) &= 1.072 \\ y(4.5) &= 1.379 & y(14.5) &= 1.169 \\ y(5.5) &= 1.566 & y(15.5) &= 1.178 \\ y(6.5) &= 1.491 & y(16.5) &= 1.116 \\ y(7.5) &= 1.239 & y(17.5) &= 1.021 \\ y(8.5) &= 0.9468 & y(18.5) &= 0.9384 \\ y(9.5) &= 0.7378 & y(19.5) &= 0.8974 \end{aligned} \tag{2.90}$$

These values give the output response at the midpoints between pairs of consecutive sampling points. Note that by varying the value of m between 0 and 1, it is possible to find the response at any intermediate points between sampling instants. The results are shown more clearly in Figure 2.8.

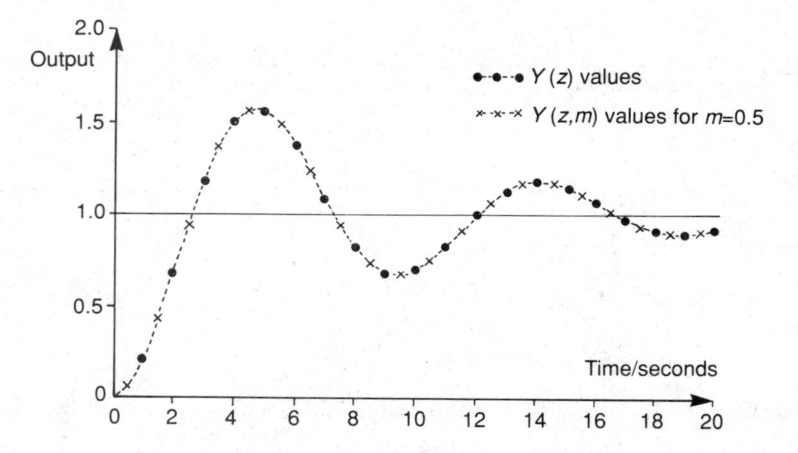

Figure 2.8 Output response using modified z-transforms

The example clearly shows the usefulness of the modified z-transform method. However, it also illustrates how even simple expressions become rather complex when modified z-transforms are used. To alleviate these practical difficulties, tables of modified z-transforms have been constructed to assist in performing such transformations (see for example Kuo [70]).

2.8 Multi-rate Sampling Systems

Analysis of the digital control systems considered so far in the book has assumed that the samplers are synchronised, and all the data in the system arises at a uniform rate. In some control applications, however, the overall system can possess loops of widely varying speeds. Hence to achieve optimised computer control performance for such systems, it may be more appropriate to sample slowly varying dynamical control loops at slow rates, and fast loops at faster rates. Also, different sample rates can be incorporated in the same control loop when input/output data via the computer occurs at different rates. For example, we may have the situation where the system output can only be measured every r^{th} sampling instant owing to sensor restrictions, but the control signal needs to be updated every step.

Analysis of Multi-rate Systems

The analysis and design of multi-rate systems is quite complicated even for simple systems (see for example Berg *et al.* [14], Flowers and Hammond [33], Jury [62], and Boykin and Frazier [16]), and we shall not go into details. A brief introduction to the area using the z-transform approach is, however, possible. The best way to start our discussion is by considering open-loop multi-rate systems, of which three basic types exist. These are as shown in

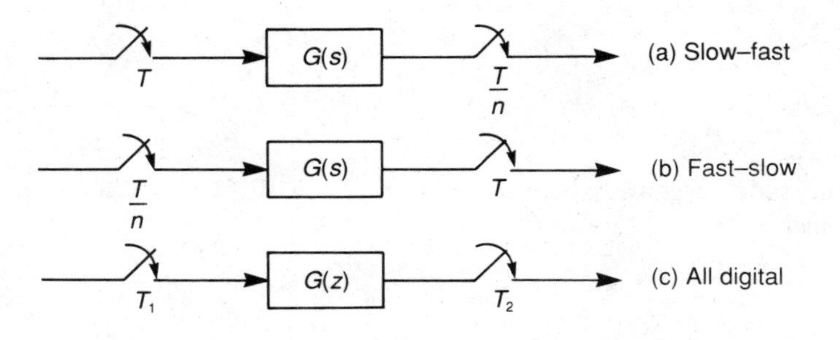

Figure 2.9 Open-loop multi-rate digital systems

Figure 2.9, where $n \geq 1$. For analytical purposes it is convenient to assume n is an integer, so that the two sample rates are integrally related (see Jury [62]; Berg *et al.* [14]). In purely digital multi-rate systems no sampling actually occurs, but for convenience we indicate the different data rates using the notation shown in Figure 2.9(c). The main methods for analysing these three forms of systems will now be presented.

Slow-Fast Multi-rate Systems

To analyse such systems, a fast-rate fictitious sampler is inserted on the input side, as shown in Figure 2.10. The subscript n means that the expressions are defined at sampling intervals of T/n, that is, at the faster rate.

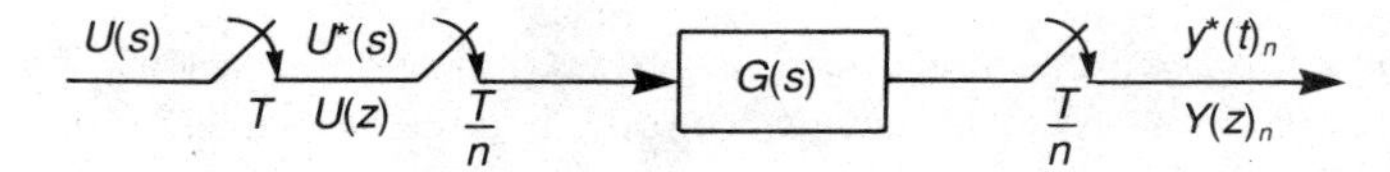

Figure 2.10 Slow-fast multi-rate system with fictitious sampler

Hence the system can be studied as if the sample rate was T/n and it can be shown that

$$Y(z)_n = G(z)_n \, U(z) \tag{2.91}$$

where

$$G(z)_n = \sum_{k=0}^{\infty} g(kT/n) \, z^{-(k/n)} \tag{2.92}$$

Fast-Slow Multi-rate Systems

Here we can introduce a fast-rate sampler in front of the slow-rate one to give rise to the situation shown in Figure 2.11. Since the output of the system

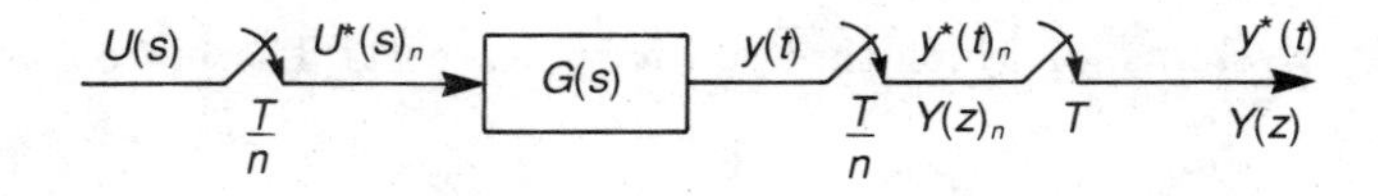

Figure 2.11 *Fast-slow multi-rate system with fictitious sampler*

is unaffected, we can analyse the system as above. Hence the z-transform of the faster output is

$$Y(z)_n = G(z)_n U(z)_n \tag{2.93}$$

where $Y(z)_n$, $G(z)_n$ and $U(z)_n$ are as above. It now remains to determine only the z-transform of the output from the slower sampler. The inverse z-transform of $Y(z)_n$ is

$$y(kT) = \frac{1}{2\pi j}\int_{\Gamma} Y(z)_n z_n^{k-1} \mathrm{d}z_n \tag{2.94}$$

where $z_n = z^{(1/n)}$. The z-transform of the system output $y(t)$ is

$$Y(z) = \sum_{m=0}^{\infty} y(mT) z^{-m} \tag{2.95}$$

Letting $m = k/n$ and using equation (2.94) in (2.95) we obtain, after a little simplification

$$Y(z) = \frac{1}{2\pi j}\int_{\Gamma} Y(z_n) \frac{1}{1 - z_n^n z^{-1}} \frac{\mathrm{d}z_n}{z_n} \tag{2.96}$$

The contour integral of equation (2.96) is carried out along the closed path Γ in the z_n-plane. Assuming the contour Γ encloses the singularities of $Y(z_n) z_n^{-1}$, and using the residue theorem, equation (2.96) can be written as

$$Y(z) = \sum \left\{ \text{Residues of } \frac{Y(z_n) z_n^{-1}}{1 - z_n^n z^{-1}} \text{ at the poles of } Y(z_n) z_n^{-1} \right\} \tag{2.97}$$

Using this equation the output response for the fast-slow multi-rate system may be determined.

All Digital Element Multi-rate Systems

In these cases, as already mentioned, no physical sampling is involved but the above techniques are still applicable.

In the slow input/fast output case shown in Figure 2.12(a), we have input data that can be represented by a train of pulses with period T seconds, and

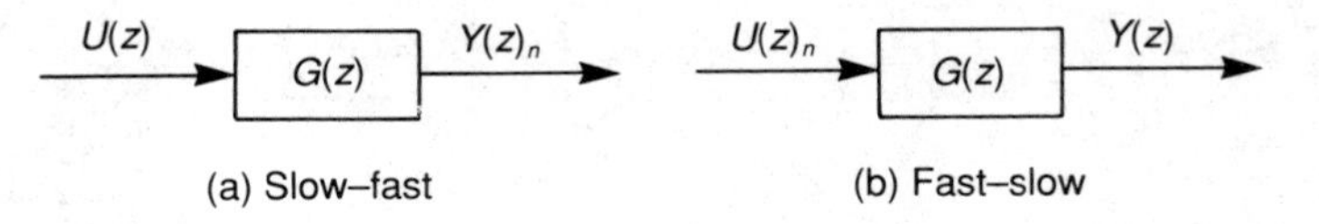

Figure 2.12 All digital multi-rate systems

output pulses occurring every T/n seconds. The input/output z-transform is written

$$Y(z)_n = G(z)_n U(z) \tag{2.98}$$

where $Y(z)_n$ and $G(z)_n$ are defined (from equation 2.92) to be of the form

$$G(z)_n = \sum_{k=0}^{\infty} g(kT/n) z^{-(k/n)} \tag{2.99}$$

For the fast-slow case shown in Figure 2.12(b), we can firstly write

$$Y(z)_n = G(z)_n U(z)_n \tag{2.100}$$

and then deduce $Y(z)$ as discussed earlier.

In this way the appropriate z-transform can be calculated to allow for the design and assessment of the system behaviour in a general multi-rate sampled system.

Closed-loop Multi-rate Systems

Many different forms of closed-loop multi-rate systems are possible, but we will consider for illustrative purposes the system shown in Figure 2.13, for which the following equations may be written

$$E(z)_n = U(z) - Y(z) - D_1(z)_n G_1(z)_n E(z)_n \tag{2.101}$$

$$Y(z)_n = D_1(z)_n G_1(z)_n G_2(z)_n E(z)_n \tag{2.102}$$

Solving for $E(z)_n$ in equation (2.101), and substituting the result in equation (2.102), we obtain

$$Y(z)_n = H(z)_n (U(z)_n - Y(z)_n) \tag{2.103}$$

where

$$H(z)_n = \frac{D_1(z)_n G_1(z)_n G_2(z)_n}{1 + D_1(z)_n G_1(z)_n} \tag{2.104}$$

It then follows that

$$Y(z) = H(z)(U(z) - Y(z)) \tag{2.105}$$

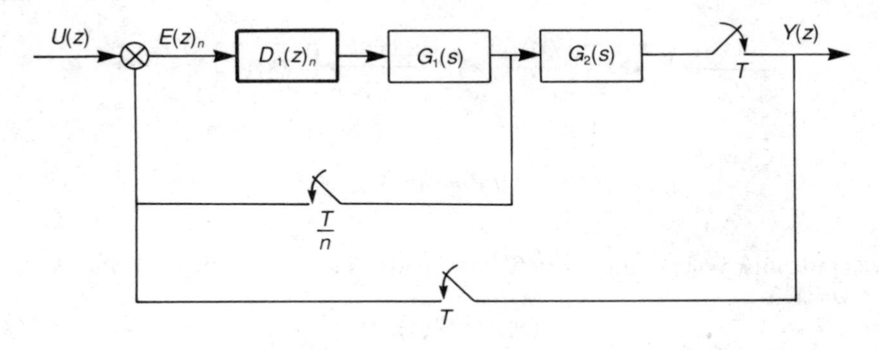

Figure 2.13 Closed-loop multi-rate system

which gives

$$Y(z) = \frac{H(z)}{1 + H(z)} U(z) \tag{2.106}$$

where $H(z)$ is obtained using the method described in the section on fast-slow multi-rate systems.

As a concluding remark on these systems it should be noted that holding devices will have different transfer functions depending upon the sampler they follow in the control loop. For example, if a ZOH comes after a sampler with period T, its Laplace transform is (see Kuo [70])

$$G_{ZOH,1} = \frac{1 - e^{-sT}}{s} \tag{2.107}$$

whereas if it follows a sampler with period T/n its transform is given by

$$G_{ZOH,2} = \frac{1 - e^{-(sT/n)}}{s}. \tag{2.108}$$

Care must therefore be taken when analysing multi-rate sampled-data systems.

2.9 Summary

The main technique for analysing discrete systems, namely the z-transform, has been introduced, together with its major characteristics. In the chapters that follow we shall make extensive use of z-domain analysis in our studies and discussions of computer control system design.

2.10 Problems

1. Converting from the continuous time domain to the z-plane, that is, $f(t) \rightarrow F(z)$, involves three steps, which are

a. sampling $f(t)$ to give $f^*(t)$;
b. Laplace-transforming $f^*(t)$ to give $F^*(s)$; and
c. letting $z = e^{sT}$ to give $F(z)$.

Use this procedure to obtain the open-form z-transforms of the following continuous time functions and then determine the equivalent z-transforms in closed-form.

(i) unit step, $u(t) = 1, \quad t \geq 0$;
(ii) $e^{-at}, \quad t \geq 0$;
(iii) $\sin \omega t$;
(iv) $\cos \omega t$.

2. Given a closed-form z-transform $Y(z)$, it is possible to find the time function $y(kT)$ by means of the following methods:
   a. power series expansion (open-form);
   b. partial fraction expansion and use of z-transform tables;
   c. inverse transform formula.

   Use these methods to inverse transform the following:

   (i) $Y(z) = \frac{2z}{(z^2-0.5z-0.5)}$
   (ii) $Y(z) = \frac{z}{(z^2+0.5)}$
   (iii) $Y(z) = \frac{4z}{(z^2-1)}$
   (iv) $Y(z) = \frac{(0.522z^2+0.361z-0.203)}{(z^3-2.347z^2+1.797z-0.449)}$

3. Consider the closed-loop system shown in Figure 2.14(a). Obtain the closed-loop discrete transfer function. If $K = 1$, calculate and describe the step response as the sampling interval is increased from 0.25 to 5 seconds.

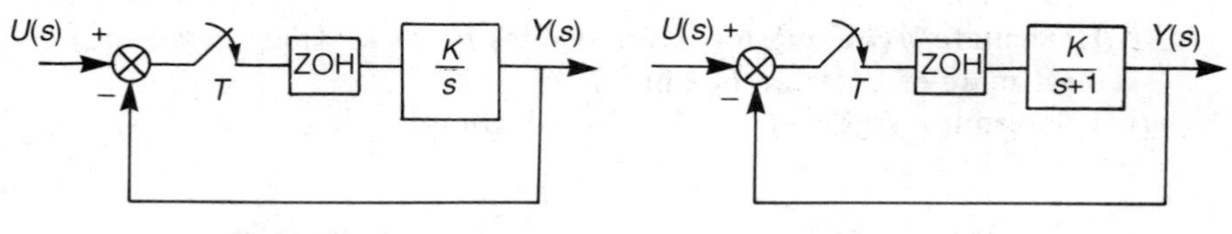

(a) Problem 3

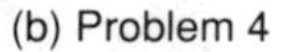
(b) Problem 4

Figure 2.14 First-order closed-loop systems

4. The system shown in Figure 2.14(b) is required to have 10% steady state error in response to a unit step input. Find the limiting values of K and T.

5. A sampled-data servo system is shown in Figure 2.15(a). Given that $T = 0.1$ s and the input is a unit step function, determine
   (i) the z-transform of the output;
   (ii) the output response of the system;
   (iii) the final value of $y(kT)$.

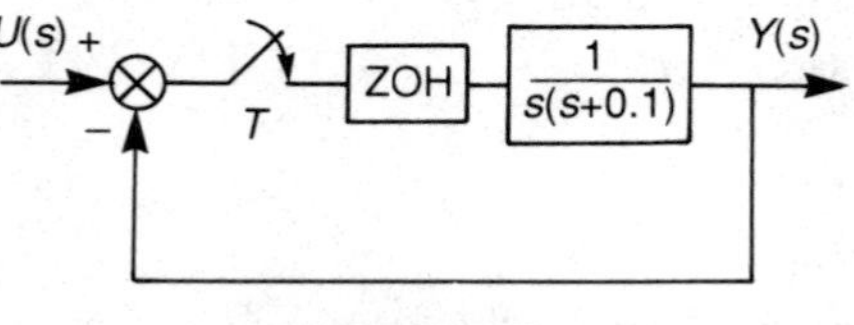

(a) Problem 5

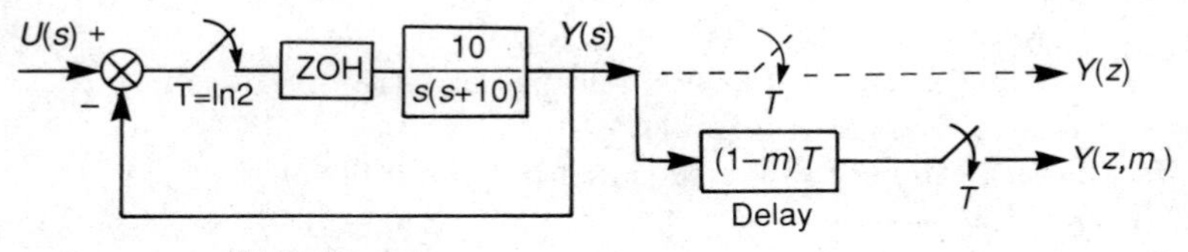

(b) Problem 6

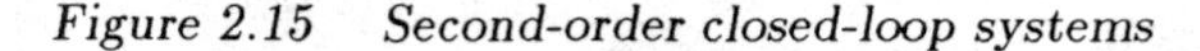

Figure 2.15 Second-order closed-loop systems

6. For the closed-loop system shown in Figure 2.15(b) determine the modified z-transform of the output $Y(z, m)$ for a unit step input. Then
   (i) calculate $Y(z)$ and check that this gives the same result as setting $m = 1$ in $Y(z, m)$;
   (ii) calculate $y(kT, m)$ for a few values of m so that an accurate estimate of $y(t)$ can be obtained;
   (iii) determine $y(kT)$ from $Y(z)$ and compare with $y(kT, m)$ for $m \neq 1$.

3 Analysis of Sampled-Data Control Systems

3.1 Introduction

In the previous chapter we introduced the z-transform and presented some of its properties for studying digital processes. We now concentrate on using the z-domain analysis techniques in the study and design of sampled data systems. Much of the discussion will be aimed at extending the results from the continuous case to the digital domain and we will assume, in the main, that the reader is familiar with the standard frequency and time domain analysis methods for the design of analogue control systems. Readers who are new to the subject can find introductions to the area in the texts by Raven [98]; D'Azzo and Houpis [25]; Ogata [88]; Kuo [71]; Van de Vegte [112]; Phillips and Harbor [94].

3.2 z-Transformation of Systems

We start by looking at the discretisation of continuous control systems where several transfer function blocks are connected together. The digitisation of a single block has already been dealt with in the previous chapter, but for the general case we need to consider two distinct cases, depending on whether samplers are present between the blocks or not.

Systems with Samplers in Cascade

The z-transform of two systems connected together with a sampler in between can be obtained by defining the signals at the various points as shown in Figure 3.1(a). Assuming that all the samplers are synchronised, we have

$$E(s) = G_1(s)U^*(s) \tag{3.1}$$

and

$$Y(s) = G_2(s)E^*(s) \tag{3.2}$$

Taking pulse transforms, that is, sampling the first equation, gives

$$E^*(s) = G^*(s)U^*(s) \tag{3.3}$$

as discussed in chapter 2, which can be substituted into equation (3.2) to give

$$Y(s) = G_2(s)G_1^*(s)U^*(s) \tag{3.4}$$

Sampling this, we have

$$Y^*(s) = G_2^*(s)G_1^*(s)U^*(s) \tag{3.5}$$

for which we can replace the * and s by z, to give the z-transform for the configuration of Figure 3.1(a) as

$$\frac{Y(z)}{U(z)} = G_1(z)G_2(z) \tag{3.6}$$

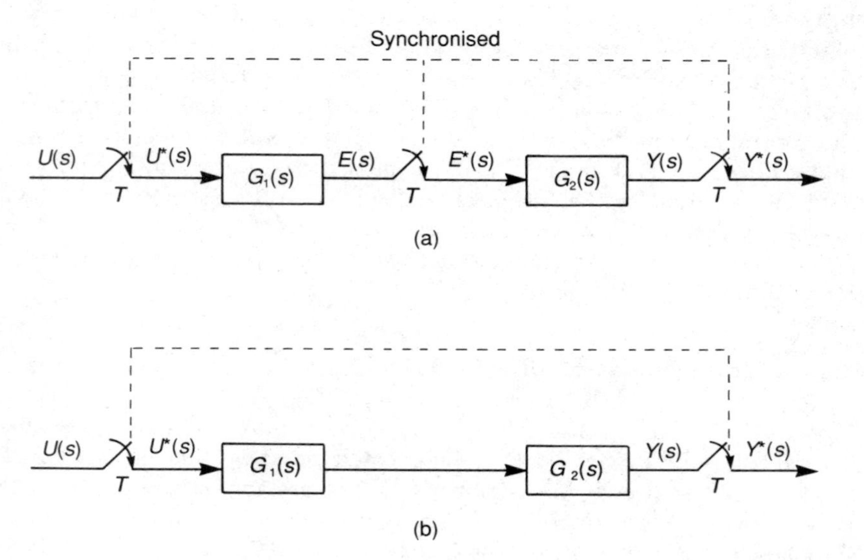

Figure 3.1 Systems in cascade

Systems in Cascade without Separating Samplers

Here the situation is as shown in Figure 3.1(b), for which we have

$$Y(s) = G_1(s)G_2(s)U^*(s) \tag{3.7}$$

Taking pulse transforms we get

$$Y^*(s) = (G_1(s)G_2(s))^*\, U^*(s) \tag{3.8}$$

from which we arrive at the z-transform as

$$\frac{Y(z)}{U(z)} = G_1G_2(z) \tag{3.9}$$

Note that $G_1G_2(z) \neq G_1(z)G_2(z)$, and so in digitising cascaded systems separated by samplers we firstly z-transform separately, before multiplying them together.

We illustrate the differences with the aid of an example. For the configuration shown in Figure 3.1, let

$$G_1(s) = \frac{1}{s} \quad \text{and} \quad G_2(s) = \frac{1}{(s+1)} \tag{3.10}$$

and let $U(s)$ be a unit step input.

1. With separating samplers we have, from tables

$$G_1(z) = \mathcal{Z}\left\{\frac{1}{s}\right\} = \frac{z}{z-1} \tag{3.11}$$

and

$$G_2(z) = \mathcal{Z}\left\{\frac{1}{s+1}\right\} = \frac{z}{z-e^{-T}} \tag{3.12}$$

and so

$$G(z) = G_1(z)G_2(z) = \frac{z^2}{(z-1)(z-e^{-T})} \tag{3.13}$$

2. Without separating samplers we have

$$G(z) = G_1G_2(z) = \mathcal{Z}\left\{\frac{1}{s(s+1)}\right\} = \frac{z(1-e^{-T})}{(z-1)(z-e^{-T})} \tag{3.14}$$

from z-transform tables.

The step responses of these are shown pictorially in Figure 3.2 (not to scale). The output from the first configuration where the blocks are separated by samplers is essentially a sequence of impulses responses, whereas the second configuration gives a sequence of step responses. The two will obviously differ as shown and thus it is important to note the locations of the samplers so that the discretisation can be carried out correctly.

Closed-Loop Systems

As discussed in chapter 1 the two main forms of closed-loop sampled data systems, shown in Figure 3.3, are sampled error, and sampled feedback. The z-transform of these is relatively straightforward to establish, using the

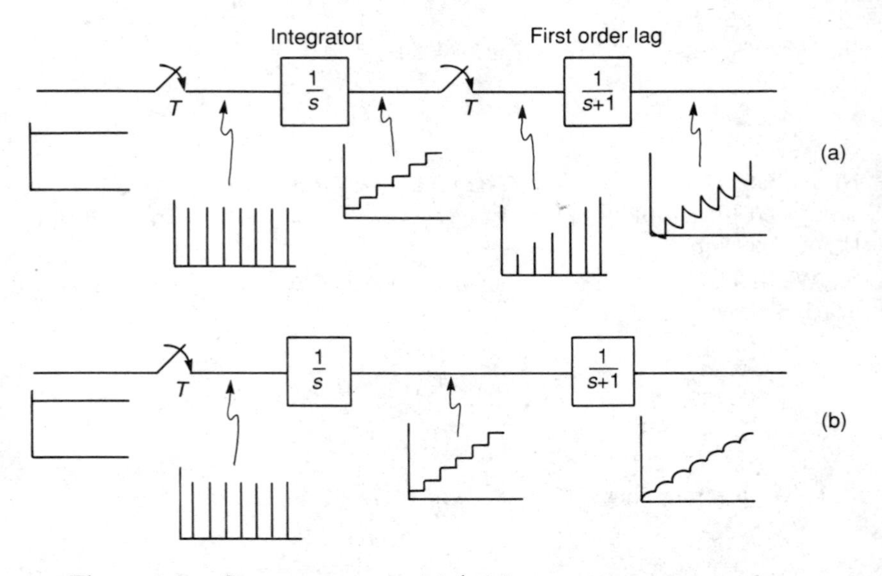

Figure 3.2 Step responses with/without separating samplers

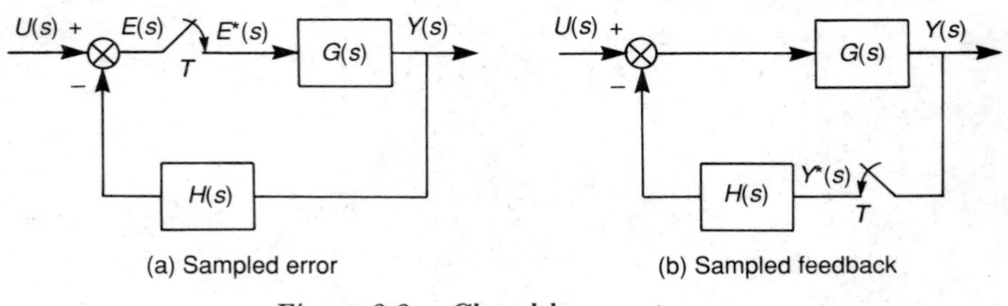

Figure 3.3 Closed-loop systems

procedure described in the previous section. For example, considering the sampled error case we have

$$Y(s) = G(s)E^*(s) \tag{3.15}$$

and

$$E(s) = U(s) - H(s)Y(s) \tag{3.16}$$

from which we obtain

$$E(s) = U(s) - H(s)G(s)E^*(s) \tag{3.17}$$

Taking pulse transforms and re-arranging we arrive at

$$E^*(s) = \frac{U^*(s)}{1 + GH^*(s)} \tag{3.18}$$

and since

$$Y^*(s) = G^*(s)E^*(s) \tag{3.19}$$

we have

$$Y^*(s) = \frac{G^*(s)U^*(s)}{1 + GH^*(s)} \tag{3.20}$$

Therefore the z-transfer function of the sampled error closed-loop system is

$$Y(z) = \frac{G(z)U(z)}{1 + GH(z)} \tag{3.21}$$

Similarly, it can easily be shown that the closed-loop transfer function for the sampled feedback case is

$$Y(z) = \frac{GU(z)}{1 + GH(z)} \tag{3.22}$$

It is therefore clear that the location of the sampler is vitally important in the analysis of discrete control systems.

The denominator of the closed-loop transfer function, when set to zero, is called the *characteristic equation* of the system. The reason for this is because the roots of this equation determine the character of the time response. The roots of the characteristic equation are also called the poles of the system. The roots of the numerator of the transfer function are called the zeros of the system.

3.3 Stability in Discrete Systems

In the design and analysis of control systems the primary objective is to ensure that the required specifications are satisfied and so it is useful to identify desirable regions in the z-plane where the poles can or cannot lie. Knowledge of the stability boundary is important in this respect, as is the relationship between the s-plane (which is well understood) and the z-plane. We start this section with a discussion on establishing the stability boundary in the z-plane.

Stability Boundary

The stable region in the z-plane can be ascertained using the properties of the s-plane together with the definition of the z-transformation. Since $z = e^{sT}$ and s is a complex number ($= \sigma + j\omega$), z also defines a complex plane with $z = e^{(\sigma + j\omega)T}$. Now

$$|\, z \,| = e^{\sigma T} \tag{3.23}$$

and

$$\angle z = \omega T \tag{3.24}$$

where $|\mu|$ is the magnitude of μ and $\angle\mu$ is the phase of μ. For $\sigma = 0$ we have continuous oscillations, or the undamped case in the s-plane. This corresponds to $|z| = 1$ and $\angle z$ changing by 2π radians (anti-clockwise) whenever ω changes by $2\pi/T$, or some multiple. Therefore the imaginary axis in the s-plane, which defines the stability boundary, is transformed into the circumference of the unit circle centred at the origin in the z-plane. For all $\sigma > 0$ (unstable in the s-plane) we have $|z| > 1$, and all corresponding points lie outside this unit circle in the z-plane. For all $\sigma < 0$ (stable in the s-plane) we have $|z| < 1$ and the points lie inside the unit circle in the z-plane. Therefore we can deduce that a digital system is stable if all its poles lie inside the unit circle centred at the origin in the z-plane, the circumference of this circle thus defining the stability boundary. A pictorial representation of this relationship between the continuous and discrete cases is shown in Figure 3.4.

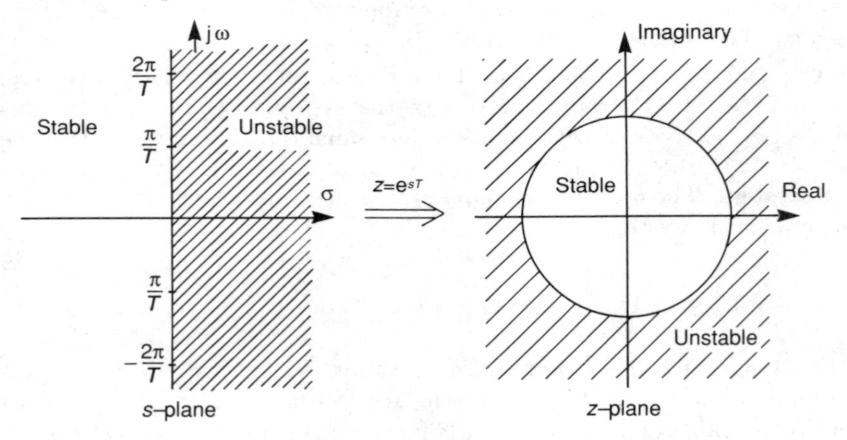

Figure 3.4 Relationship between the s- and z-planes

Having established the stable and unstable regions in the z-plane it is worth analysing the numerical integration approximations discussed in chapter 2, to determine how well the methods perform and why. It was seen that starting with a stable continuous system, the forward rectangular method yielded an unstable digital realisation. Also, the other two methods considered, namely the backward rectangular rule and Tustin's rule, gave differing degrees of accuracies in the approximations for the same system. To understand the reasons for these results each of the three transformations may be viewed graphically to determine how the stability region in the s-plane maps into the z-plane. For this purpose we must solve the three relations for z in terms of s. It is straightforward to deduce that the following relations hold:

(i)	forward rectangular rule	$z = 1 + Ts$,
(ii)	backward rectangular rule	$z = 1/(1 - Ts)$,
(iii)	Tustin's rule	$z = (1 + Ts/2)/(1 - Ts/2)$.

The shaded regions show the transformations of the left half s-plane to the z-plane when different approximations are used

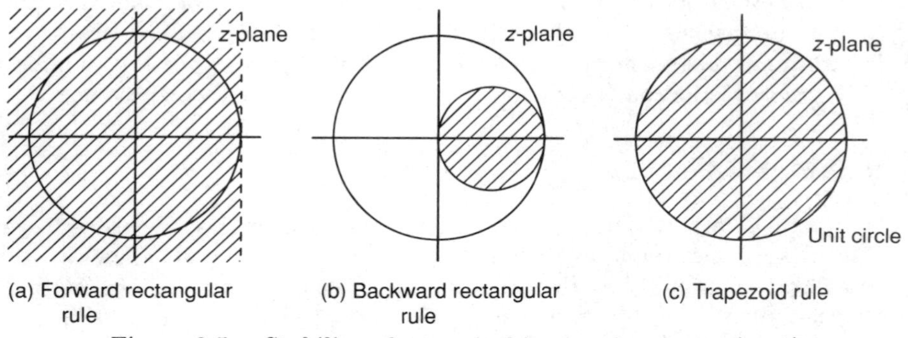

Figure 3.5 Stability of numerical integration approximations

If we set $s = j\omega$ in these equations we can obtain the stability boundaries in the z-plane for each case, as shown in Figure 3.5. To demonstrate that the backward rule gives the circular stability region as shown, 1/2 is added to and subtracted from (ii) to yield

$$z = \frac{1}{2} - \frac{1}{2}\left(\frac{1 + Ts}{1 - Ts}\right) \tag{3.25}$$

When $s = j\omega$ the magnitude of $(z - 1/2)$ is constant and equals

$$\mid z - 1/2 \mid = 1/2 \tag{3.26}$$

and the curve is thus a circle as shown in Figure 3.5(b).

Since the unit circle gives the stability boundary in the z-plane, it is obvious from Figure 3.5 that the forward rectangular rule can give an unstable transformation for a stable continuous transfer function (as illustrated by the example considered in chapter 2). The other two transformations will always give stable digital realisations for stable continuous systems. It is also important to note that, since Tustin's rule maps the stable region of the s-plane exactly into the stable region of the z-plane, it gives rise to better approximations as seen in chapter 2. However this does not imply perfect results because the entire $j\omega$-axis of the s-plane is squeezed into the 2π length of the unit circle in the z-plane. Clearly a distortion takes place in the mapping in spite of the agreement of the stability regions. Steps can be taken to attempt to correct for this distortion, see for example Franklin and Powell [35]. Tustin's transformation is further discussed in section 3.6 since this is the method most commonly used in practice.

Since the stability boundary in the z-plane is the unit circle centred at the origin, it appears straightforward to determine whether a particular system is stable or not, simply by determining the location of its poles. In practice the situation can be rather more complicated. The main difficulty is that for large order polynomials the roots are not readily available, and use of computer packages is necessary for the factorisation. Such difficulties have led to the development of methods that do not require explicit evaluation of the roots of the characteristic polynomials, and we discuss these next.

Routh's Method

It is well known that for the continuous case the method developed by Routh (see for example D'Azzo and Houpis [25]; Raven [98]) allows us to assess system stability by the construction of a simple array, known as the Routh array. Here, by calculating various determinants, it is possible to determine how many system poles are in the unstable right-half s-plane.

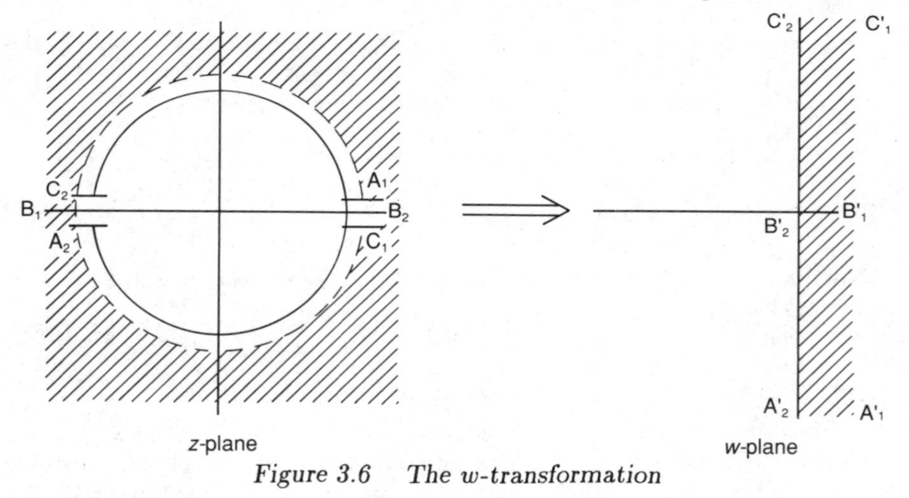

Figure 3.6 The w-transformation

The Routh method can be extended to discrete systems if the different stability boundary in the z-domain can be modified, that is, the unit circle at the origin in the z-domain needs to be converted to the imaginary axis. A bilinear transformation w achieves this by converting the interior of the unit circle (in the z-plane) into the left half plane (in the w-plane), as shown in Figure 3.6. The transformation necessary for this is defined by

$$w = \frac{z+1}{z-1} \tag{3.27}$$

which gives

$$z = \frac{w+1}{w-1} \tag{3.28}$$

As can be seen from the curve $A_1B_1C_1$ in Figure 3.6, the point $z = 1$ is undefined in the w-plane using this definition. For systems with poles at this point it is possible to use an alternative definition for w given by

$$w = \frac{z-1}{z+1} \tag{3.29}$$

which gives

$$z = \frac{1+w}{1-w} \tag{3.30}$$

so that the poles at $z = 1$ can be transformed into the w-plane. Using this transformation, curve $A_2B_2C_2$, in Figure 3.6 shows that the undefined point is now at $z = -1$. Hence, the appropriate transformation can be employed to convert a discrete characteristic polynomial, $F(z)$, into an equivalent polynomial, $F(w)$, in the w-plane. To assist in this transformation there are conversion tables (see, for example Kuo [70]). Now, since the w-plane is similar to the s-plane, it is possible to apply the standard Routh method to the characteristic equation in w for assessing system stability.

We illustrate the technique by considering the simple example of a discrete system whose characteristic equation is

$$F(z) = 3z^4 + z^3 - z^2 - 2z + 1 = 0 \tag{3.31}$$

To apply the Routh method we firstly apply the transformation

$$z = \frac{w+1}{w-1} \tag{3.32}$$

to give $F(w)$, where

$$F(w) = 2w^4 + 14w^3 + 26w^2 + 2w + 4 = 0 \tag{3.33}$$

Forming the Routh array we obtain Table 3.1, to which the Routh stability result can be applied. This states that the system has no unstable roots if all the numbers in the first column have the same sign, and also that none of the numbers vanishes. In our example there are two sign changes, and so there are two unstable poles. In fact, $F(z)$ and $F(w)$ have the following roots:

$F(z)$: $z = -0.74 \pm j0.69$ (unstable poles), or $0.57 \pm j0.08$.

$F(w)$: $w = 0.003 \pm j0.39$ (unstable poles), or $-3.5 \pm j0.79$.

Table 3.1 Routh array for example

w^4	2	26	4
w^3	14	2	
w^2	25.7	4	
w	-0.2		
w^0	4		

Jury's Method

An alternative, array based, stability assessment method that addresses the discrete characteristic polynomial directly was developed by Jury and Blanchard [63]; Jury [61]. In this case, though there is no need to transform to another domain, the array construction and assessment conditions are more involved. Consider the characteristic polynomial

$$F(z) = a_n z^n + a_{n-1} z^{n-1} + \cdots + a_1 z + a_0 = 0 \tag{3.34}$$

where $a_0, a_1, \ldots, a_n$ are real coefficients and $a_n > 0$. To construct Jury's table, the a_i coefficients are inserted on the first two rows of the array and are then used to calculate the complete array shown in Table 3.2.

Table 3.2 Form of Jury's array

Row	z^0	z^1	z^2	$\cdots$	z^{n-2}	z^{n-1}	z^n
1	a_0	a_1	a_2	$\cdots$	a_{n-2}	a_{n-1}	a_n
2	a_n	a_{n-1}	a_{n-2}	$\cdots$	a_2	a_1	a_0
3	b_0	b_1	b_2	$\cdots$	b_{n-2}	b_{n-1}	
4	b_{n-1}	b_{n-2}	b_{n-3}	$\cdots$	b_1	b_0	
5	c_0	c_1	c_2	$\cdots$	c_{n-2}		
6	c_{n-2}	c_{n-3}	c_{n-4}	$\cdots$	c_0		
$\vdots$	$\vdots$	$\vdots$	$\vdots$	$\vdots$			
$2n-5$	r_0	r_1	r_2	r_3			
$2n-4$	r_3	r_2	r_1	r_0			
$2n-3$	s_0	s_1	s_2				
$2n-2$	s_2	s_1	s_0				

In forming the array we have

$$b_k = \begin{vmatrix} a_0 & a_{n-k} \\ a_n & a_k \end{vmatrix}, \; c_i = \begin{vmatrix} b_0 & b_{n-1-i} \\ b_{n-1} & b_i \end{vmatrix}, \; \begin{array}{l} \text{for } k = 0, 1, \ldots, n-1, \\ \text{for } i = 0, 1, \ldots, n-2, \text{ etc.} \end{array} \tag{3.35}$$

and $| \, M \, |$ represents the determinant of the matrix M. Note that the elements of the $(2k+2)^{\text{th}}$ rows for $k = 0, 1, 2, \ldots, n-2$, consist of the coefficients in the $(2k+1)^{\text{th}}$ row written in reverse order. Having constructed the above table we can apply Jury's stability test which states the necessary and sufficient conditions for $F(z)$ to have no roots on, or outside, the unit circle. The conditions are as follows:

$$\begin{array}{llcll} \text{(i)} & [F(z)]_{z=1} & > & 0 & \\ & & & & \\ \text{(ii)} & [F(z)]_{z=-1} & > & 0 & \text{for } n \text{ even} \\ & & < & 0 & \text{for } n \text{ odd} \\ & & & & \\ \text{(iii)} & | a_0 | & < & | a_n | & \\ & | b_0 | & > & | b_{n-1} | & \\ & | c_0 | & > & | c_{n-2} | & \\ & \vdots & \vdots & \vdots & \\ & | s_0 | & > & | s_2 | & \end{array} \tag{3.36}$$

As an example, we again consider the system defined by the characteristic polynomial

$$F(z) = 3z^4 + z^3 - z^2 - 2z + 1 = 0 \tag{3.37}$$

but now assess its stability by Jury's method.

Since conditions (i) and (ii) do not depend on the table, it is advisable to check them before constructing the array.

- $f(1) = 2 > 0$, and so condition (i) is satisfied.
- $f(-1) = 4 > 0, \quad n = 4$, is even, and so condition (ii) is also satisfied.

We can therefore construct Jury's array as shown in Table 3.3, from which we have

$$\begin{array}{lllll} | a_0 | = 1, & | a_4 | = 3 & \Longrightarrow & | a_0 | < | a_4 | & \text{satisfied} \\ | b_0 | = 8, & | b_3 | = 7 & \Longrightarrow & | b_0 | > | b_3 | & \text{satisfied} \\ | c_0 | = 15, & | c_2 | = 19 & \Longrightarrow & | c_0 | \not> | c_2 | & \text{is } \textit{not} \text{ satisfied.} \end{array} \tag{3.38}$$

and so we can deduce that the system is unstable as discovered earlier.

Table 3.3 *Jury's array for example*

Row		z^0	z^1	z^2	z^3	z^4
1	a_0	1	−2	−1	1	3
2	a_4	3	1	−1	−2	1
3	b_0	−8	−5	2	7	
4	b_3	7	2	−5	−8	
5	c_0	15	26	19		
6	c_2	19	26	15		

Raible's Method

A simplification for the construction of Jury's array by hand was suggested by Raible [97]. Considering

$$F(z) = a_n z^n + a_{n-1} z^{n-1} + \cdots + a_1 z + a_0 = 0 \tag{3.39}$$

with the a_i values real and $a_n > 0$ as before, the Raible table is constructed as shown in Table 3.4, where

$$\begin{aligned} &k_a = \tfrac{a_0}{a_n}, \quad k_b = \tfrac{b_{n-1}}{b_0}, && k_c = \tfrac{c_{n-2}}{c_0}, \ldots, \quad k_q = \tfrac{q_1}{q_0}, \\ &b_i = a_{n-i} - k_a a_i && \text{for } i = 0, 1, 2, \ldots, n-1 \\ &c_j = b_j - k_b b_{n-j-1} && \text{for } j = 0, 1, 2, \ldots, n-2, \text{ etc.} \end{aligned} \tag{3.40}$$

Table 3.4 *Form of Raible's array*

z^n	z^{n-1}	z^{n-2}	$\cdots$	z^2	z^1	z^0	k
a_n	a_{n-1}	a_{n-2}	$\cdots$	a_2	a_1	a_0	k_a
b_0	b_1	b_2	$\cdots$	b_{n-2}	b_{n-1}		k_b
c_0	c_1	c_2	$\cdots$	c_{n-2}			k_c
$\vdots$	$\vdots$	$\vdots$					$\vdots$
p_0	p_1	p_2					k_p
q_0	q_1						k_q
r_0							

We then have the following result:

Raible's Criterion

The number of roots of $F(z) = 0$ inside the unit circle is equal to the number of positive calculated elements in the first column

$$[b_0,\ c_0,\ d_0,\ \ldots,\ q_0,\ r_0]^T \tag{3.41}$$

or equivalently, the number of negative calculated elements in the first column gives the number of roots outside the unit circle.

Again, to illustrate the method, we consider the system which has the characteristic polynomial

$$F(z) = 3z^4 + z^3 - z^2 - 2z + 1 = 0 \tag{3.42}$$

Forming Raible's table by hand we have the result shown in Table 3.5.

Table 3.5 Raible's array for example

	z^4	z^3	z^2	z^1	z^0	k
$a's$	3	1	−1	−2	1	0.33
$-k_a a_i$	−0.33	0.67	0.33	−0.33		
$b's$	**2.67**	1.67	−0.67	−2.33		−0.87
$-k_b b_{3-j}$	−2.03	−0.58	1.46			
$c's$	**0.64**	1.09	0.79			1.23
$-k_c c_{2-j}$	−0.97	1.34				
$d's$	**−0.33**	−0.25				0.76
$-k_d d_{1-j}$	0.19					
e	**−0.14**					

Note that in the calculation of the array we have inserted intermediate rows (for example $-k_a a_i$) to assist the hand calculations. The calculated first column is $[2.67,\ 0.64,\ -0.33,\ -0.14\ ]^T$ from which we can deduce that $F(z)$ has two roots inside the unit circle, and two outside (hence the system is unstable).

Singular Cases

In the above discussion on the methods of Jury and Raible we have assumed that the calculations can be continued to the last row on the table. In certain cases some or all of the row elements are zero and the tabulations end prematurely. These situations are referred to as singular cases and are caused by roots lying on the stability boundary, that is, on the unit circle. The problems posed by singular cases can be eliminated by expanding and contracting the stability boundary so that the roots no longer lie on it. The transformation suitable for this is

$$z = (1 + \varepsilon)z \tag{3.43}$$

where ε is a small real number. When ε is a positive number the unit circle radius is expanded to $1+\varepsilon$ and for negative ε the circle is contracted. By checking the number of roots inside and outside the stability boundary as this expansion and contraction occurs, it is possible to determine how many roots lie inside, outside and actually on the unit circle. The transformation is relatively straightforward to apply, since the new polynomial coefficients can be approximated as follows:

> For z^n terms, the $z = (1 \pm \varepsilon)z$ transformation can be approximated to $(1 \pm n\varepsilon)z^n$ by use of the binomial series expansion since ε is small. Hence the coefficients of the z^n terms needs to be multiplied by $(1 \pm n\varepsilon)$, etc.

The method is best illustrated by an example, where we assess the stability and the root locations of the system whose characteristic equation is

$$F(z) = z^3 + 3.6z^2 + 3.88z + 2.4 = 0 \tag{3.44}$$

We will concentrate on Raible's method, although similar results can be obtained via Jury's approach. Forming Raible's array as usual, we obtain Table 3.6 which shows that we have a singular situation as the tabulation has terminated prematurely.

Table 3.6 Raible's array for singular example

	z^3	z^2	z	z^0	k
a_3	1	3.6	3.88	2.4	2.4
$-k_a a_i$	-5.76	-9.312	-8.64		
b_2	$\mathbf{-4.76}$	-5.712	-4.76		1
$-k_b b_{2-j}$	-4.67	-5.712			
c_1	**0**	0			

It is therefore necessary to transform $z \to (1+\varepsilon)z$ to be able to vary the stability boundary. The transformed characteristic polynomial is

$$F(1+\varepsilon z) \approx (1+3\varepsilon)z^3 + 3.6(1+2\varepsilon)z^2 + 3.88(1+\varepsilon)z + 2.4 = 0 \tag{3.45}$$

where we have ignored all higher order terms since ε is small. The Raible array for equation (3.45) is then obtained as shown in Table 3.7. Looking at the calculated elements in the first column we have:

$$\begin{array}{ll} \text{when } \varepsilon > 0 \implies b_2 \text{ is } -\text{ve}, \; c_1 \text{ is } +\text{ve}, \; d_0 \text{ is } +\text{ve} \\ \text{when } \varepsilon < 0 \implies b_2 \text{ is } -\text{ve}, \; c_1 \text{ is } -ve, \; d_0 \text{ is } -\text{ve} \end{array} \tag{3.46}$$

Hence the system is unstable with one root in the unstable region and two roots on the stability boundary. In fact $F(z)$ has roots at $z = -2.4$ and $-0.6 \pm j0.8$.

Table 3.7 Modified Raible's array for singular example

	z^3	z^2	z	z^0	k
a_3	$1+3\varepsilon$	$3.6+7.2\varepsilon$	$3.88+3.88\varepsilon$	2.4	$\frac{2.4}{1+3\varepsilon}$
$-k_a a_i$	$\frac{-5.76}{1+3\varepsilon}$	$\frac{-9.312-9.31\varepsilon}{1+3\varepsilon}$	$\frac{-8.64-17.28\varepsilon}{1+3\varepsilon}$		
b_2	$\frac{-4.76+6\varepsilon}{1+3\varepsilon}$	$\frac{-5.71+8.688\varepsilon}{1+3\varepsilon}$	$\frac{-4.76-1.76\varepsilon}{1+3\varepsilon}$		$\frac{4.76+1.76\varepsilon}{4.76-6\varepsilon}$
$-k_b b_{2-j}$	$\frac{22.66+16.67\varepsilon}{(1+3\varepsilon)(4.76-6\varepsilon)}$	$\frac{27.19-31.3\varepsilon}{(1+\varepsilon)(4.76-6\varepsilon)}$			
c_1	$\frac{73.88\varepsilon}{(1+3\varepsilon)(4.76-6\varepsilon)}$	$\frac{44.3\varepsilon}{(1+\varepsilon)(4.76-6\varepsilon)}$			0.6
	$\frac{-26.56\varepsilon}{(1+3\varepsilon)(4.76-6\varepsilon)}$				
d_0	$\frac{47.32\varepsilon}{(1+3\varepsilon)(4.76-6\varepsilon)}$				

3.4 Time Domain Analysis

Having spent some time on the stability aspects for discrete systems, we now look at the main methods for analysing the transient and steady state behaviour of these systems. As for the continuous case, the time response of digital systems can be characterised by such items as the overshoot, rise time t_r, delay time t_d, settling time t_s, time to first peak t_p, damping ratio ζ, and natural undamped frequency ω_n. By applying test signals such as step inputs, the system performance can be determined by studying the output responses. A typical unit step response illustrating some of the time domain performance criteria is shown in Figure 3.7. Again as in the continuous case, using standard second-order system responses (see appendix B), we can categorise systems by, for example, their dominant roots. Notice that although we are dealing with digital systems, the output is usually still a continuous variable and so $t_d, t_s, t_r, \zeta, \omega_n$ can be defined as in the purely continuous case. However, when the z-transform is used, the time responses are only defined at the sampling instants, and so care must be taken to ensure the accuracy and validity of the discrete analysis. Essentially such problems can be avoided by appropriate selection of the sampling interval.

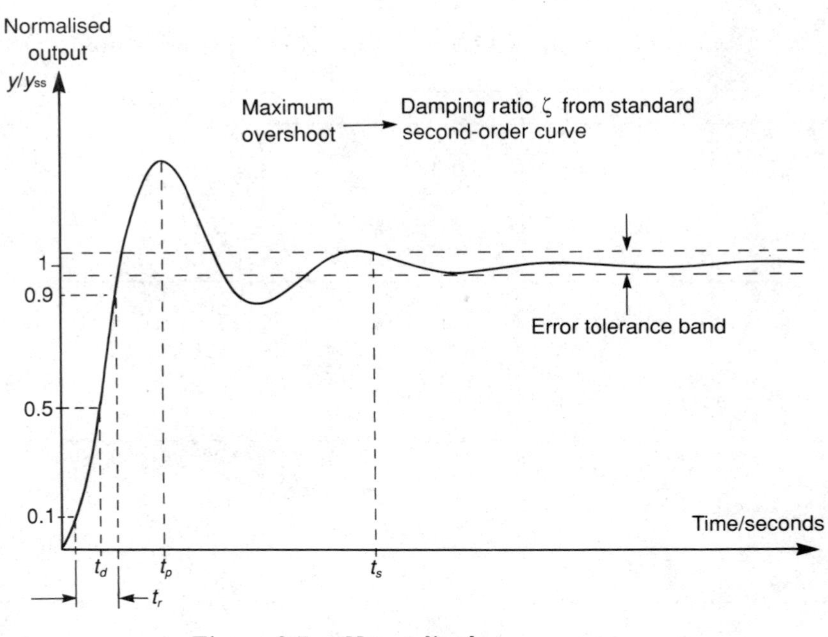

Figure 3.7 Normalised step response

For a fuller discussion on this and other aspects of time domain analysis for digital systems, see Kuo [70]; Franklin and Powell [35]; Houpis and Lamont [46]; Ogata [89].

3.5 Root Locus Analysis

The root locus method is a powerful method for the study of continuous systems. Well defined rules (see for example D'Azzo and Houpis [25]) for drawing the root locus exist, so that the migration of closed-loop poles, as a function of gain parameter variation, can be obtained. This ability assists greatly in the design of controllers so that the required damping ratios, speed of response, etc. are satisfied. It turns out that we can extend the method to the z-plane quite easily. Consider the closed-loop discrete system shown in Figure 3.8 which has a closed-loop transfer function

$$\frac{Y(z)}{U(z)} = \frac{kG(z)}{1 + kG(z)H(z)} \tag{3.47}$$

It is well known that the characteristic equation

$$1 + kG(z)H(z) = 0 \tag{3.48}$$

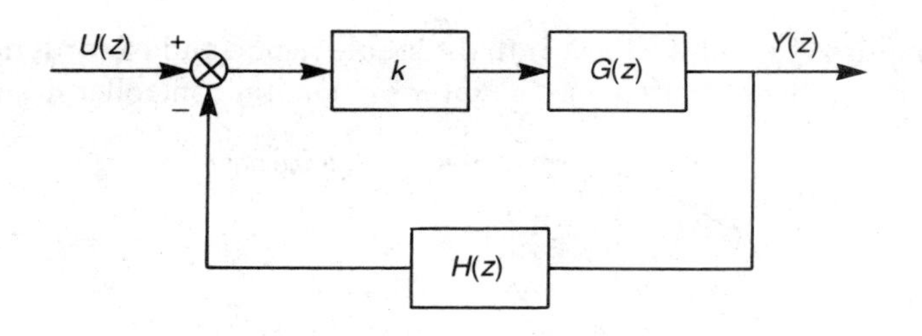

Figure 3.8 *Closed-loop discrete system*

gives the closed-loop poles. As the gain k varies, the root locus in the z-plane is obtained following the standard (continuous domain) rules, where the locus starts at the open-loop poles of the system and terminates at the open-loop zeros. However, the interpretation of the locus is different from the continuous case, because of the different stability boundary as discussed in section 3.3.

A simple example is presented next to illustrate the method. Consider a double integrator system shown in Figure 3.9(a), which can represent a single degree of freedom motion of a rigid-body satellite system. This system is further studied in chapter 4. When z-transformed using tables, the system becomes as shown in Figure 3.9(b).

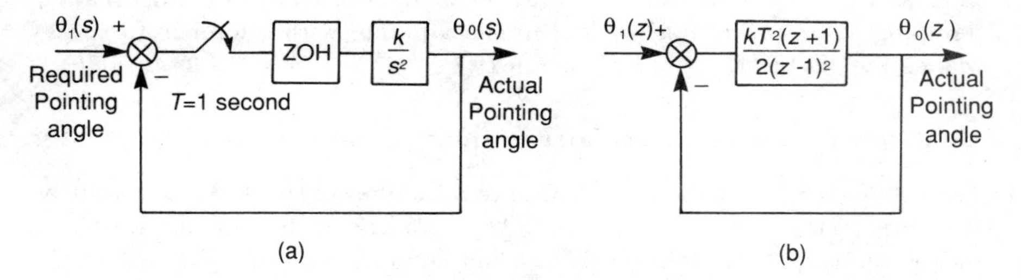

Figure 3.9 *Computer control of a rigid-body satellite*

Hence the open-loop discrete system has two poles at $z = 1$ and one zero at $z = -1$. Since $T = 1$ s the characteristic equation is equal to

$$1 + \frac{k(z+1)}{2(z-1)^2} = 0 \tag{3.49}$$

so that the closed-loop poles are given by the roots of

$$z^2 + z\,(k/2 - 2) + 1 + k/2 = 0 \tag{3.50}$$

As k varies, the root locus shown in Figure 3.10 is obtained, from which we can see that the system is unstable for all k. Some form of compensation is therefore required to pull the locus into the stable region and towards desired locations within the unit circle (see chapter 4). Digital controller design (in

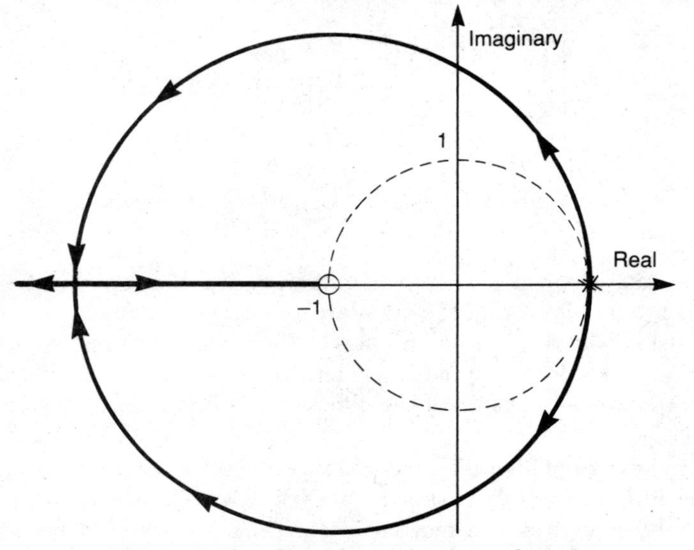

Figure 3.10 Root locus in the z-domain

particular, via the root locus method) can be facilitated by drawing constant damping ratio and frequency loci in the z-plane, with reference to the s-plane. We discuss how this is done next.

Constant Damping Ratio Loci in the z-plane

We wish to transform constant damping ratio lines in the s-plane to equivalent lines in the z-plane. Consider the $\zeta = 0.5$ line in the s-plane, as shown in Figure 3.11(a). Points on this line can be represented as $(-0.577 + j)\alpha$, where α is a real scalar with $\alpha > 0$. Then, by definition of the z-transform, these points transform to

$$z = \exp\{(-0.577 + j)\,\alpha T\} = \exp\{(-0.577 + j)\,\alpha 2\pi/\omega_s\} \tag{3.51}$$

In the s-plane, as $\alpha \to \infty$, we move to ∞ on the $\zeta = 0.5$ line. The corresponding points in the z-plane can be obtained by using the above transformation. For example, when $\alpha = 0$, $z = 1\angle 0°$, when $\alpha = \omega_s/4$, $z = 0.39\angle 90°$, and when $\alpha = \omega_s/2$, $z = 0.153\angle 180°$. The collection of all such points in the z-plane gives the $\zeta = 0.5$ curve as shown in Figure 3.11(b).

Rise-time Considerations

The rise-time t_r is defined to be the time taken for the system output to rise from 10% to 90% of its final value. As can be seen from the standard second-

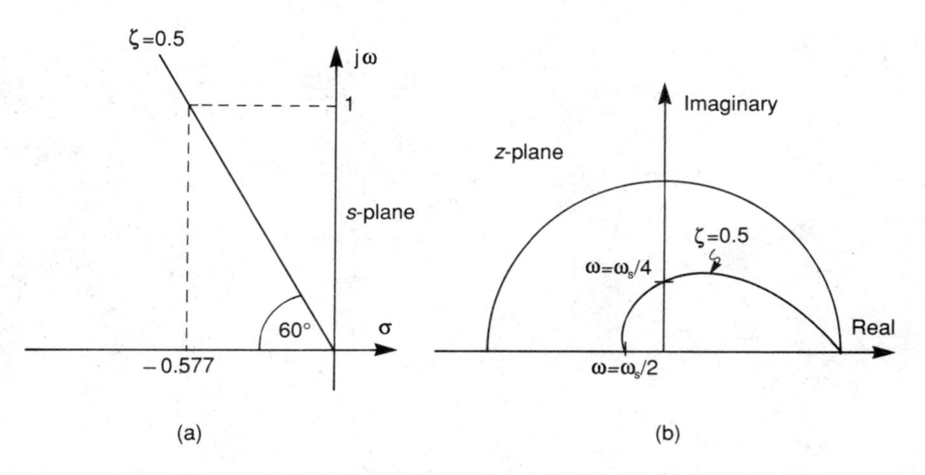

Figure 3.11 Transformation of constant ζ lines

order system results in appendix B, t_r is dependent upon the damping ratio, ζ, and the undamped natural frequency of the system, ω_n. For example, if $\zeta = 0.5, \quad \omega_n t_r \approx 2.5$ and hence

$$\omega_n \geq \frac{2.5}{t_r} \tag{3.52}$$

Therefore the natural frequency of the system needs to be greater than this expression in order to achieve the rise-time requirements. In the s-plane this corresponds to radiating out from the origin at the appropriate slope, for different ζ, as shown in Figure 3.12(a). The z-plane results are again obtained by the transformation $z = \exp\{(\sigma + j\omega)\alpha T\}$. Looking at the $\zeta = 0$ line (on the imaginary axis in the s-plane), $\sigma = 0$ and letting $\omega = 1$ we have

$$z = \exp\{j2\pi\alpha/\omega_s\} \tag{3.53}$$

that is, z starts from the point $1\angle 0°$ and lies on the unit circle moving in an anti-clockwise direction as $\alpha \to \infty$. For $\zeta = 1$ (the real axis in the s-plane), $\omega = 0$ and letting $\sigma = 1$ we have $z = \exp\{\alpha 2\pi/\omega_s\}$, that is, z decreases to zero on the real line from the point z=1 as α decreases to $-\infty$.

The other constant ζ lines between 0 and 1 "fill up" the unit circle in the z-plane. In fact it can be shown that, in the z-plane, the lines of constant ω_n are lines drawn at right angles to constant ζ lines (see for example Kuo [70]; Leigh [74]). The rise-time condition is therefore satisfied for the discrete case if the roots do not lie in the shaded area, as shown in Figure 3.12(b) where ω_n is increasing as we go anti-clockwise.

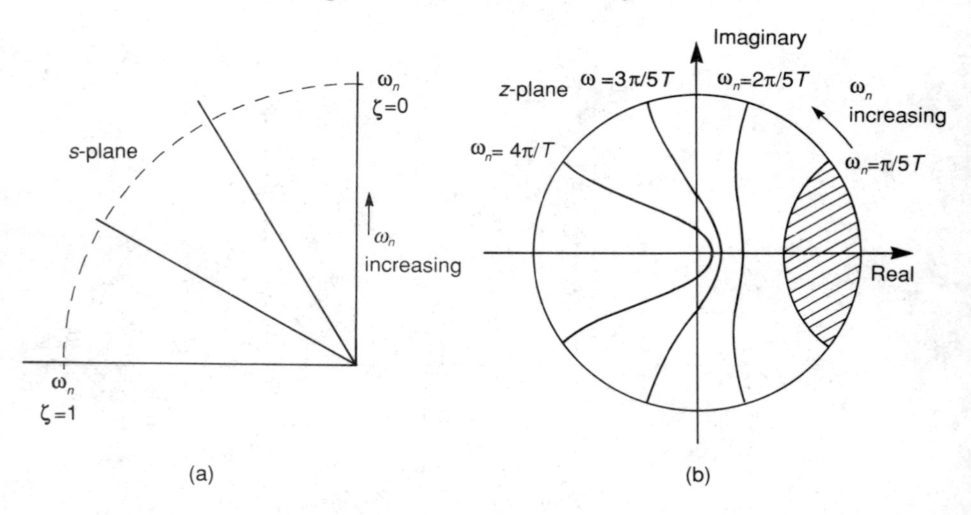

Figure 3.12 Rise-time transformation

Settling Time Considerations

The settling time is defined to be the time taken for the output to settle within some error band. It is straightforward to show (see D'Azzo and Houpis [25]) that the output response of a standard second-order system is given by

$$y(t) = 1 - e^{(-\zeta\omega_n t)} \cos(\omega_d t + \phi) \tag{3.54}$$

where ω_d is the damped frequency of the system, and ϕ is the phase shift. The oscillatory behaviour is contained in an envelope defined by the exponential whose time constant is given by the real part of the root location in the s-plane. If the error tolerance is 1%, we require the exponential term to have decayed to this level for settling to have occurred, that is

$$e^{-\zeta\omega_n t_s} \leq 0.01 \tag{3.55}$$

This gives

$$\zeta\omega_n \geq \frac{4.6}{t_s} \tag{3.56}$$

Hence we have

$$\text{the real part of } \{s\} \leq -4.6/t_s \tag{3.57}$$

In the z-domain we have $|\, z \,| = e^{-\zeta\omega_n T}$, which is the distance from the origin. Therefore, for the digital system to satisfy the settling time requirements, all roots must lie within a circle whose radius is $\exp(-4.6T/t_s)$, as shown in Figure 3.13. The above results are incorporated onto special z-domain root locus paper, shown in Figure 3.14, on which designs can be readily carried

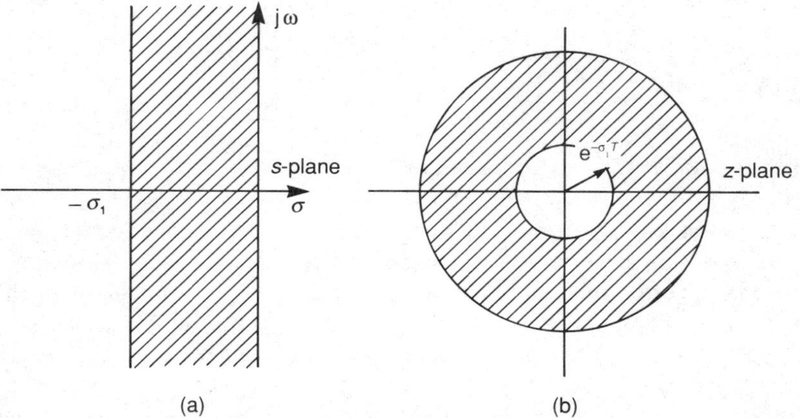

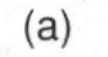

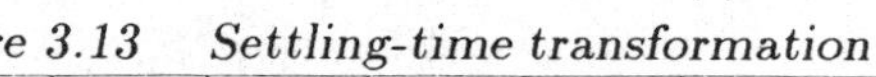
Figure 3.13 Settling-time transformation

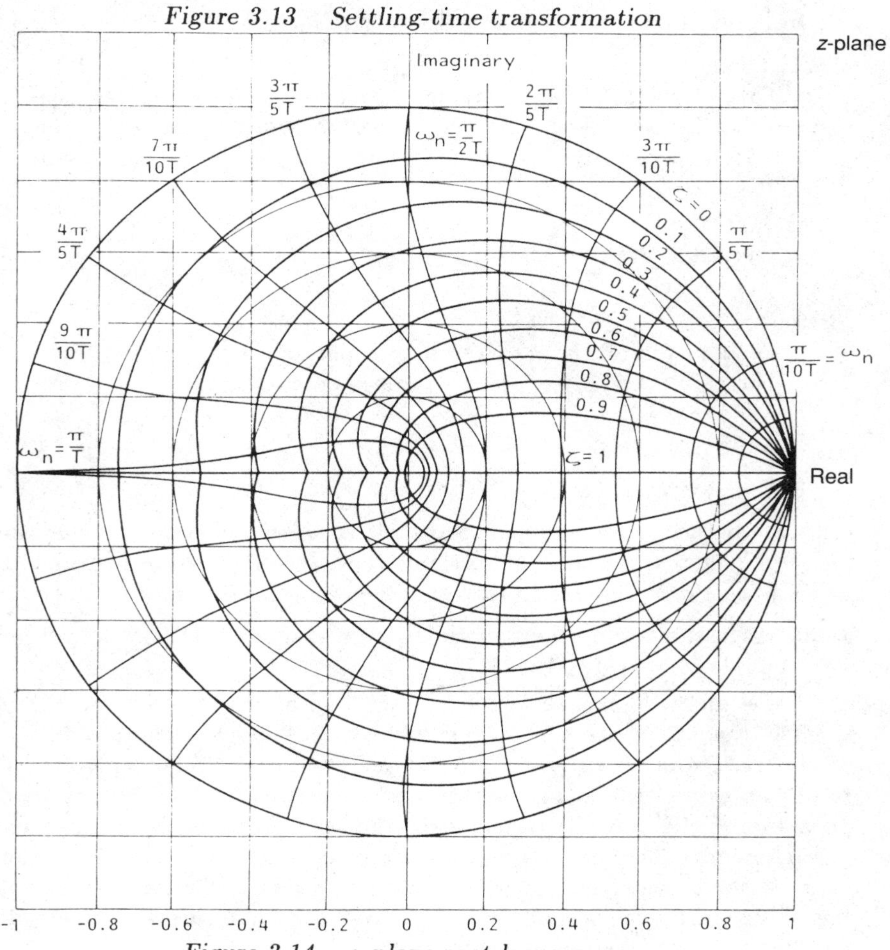

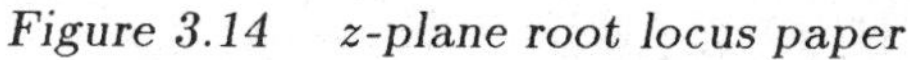
Figure 3.14 z-plane root locus paper

out. An example of its use will be given in chapter 4 when we discuss controller design.

3.6 Frequency Domain Analysis

When analysis is performed in the frequency domain, the system performance is given in terms of the steady state response when the input is sinusoidal. Terms such as phase margin, gain margin, and bandwidth are used in specifying the performance (see for example D'Azzo and Houpis [25]). The frequency response of a continuous system with open-loop transfer function, $G(s)$, is obtained by letting $s = j\omega$ and then observing the behaviour of $G(j\omega)$ as ω is increased from 0 to ∞. Several different methods for studying the frequency data exist. These include:

Nyquist Plot

This plots the $G(j\omega)$ data in the complex plane, as cartesian coordinates, as $\omega \to \infty$. The Nyquist stability criterion states that encirclement of the $(-1, 0)$ point implies that the closed-loop system will be unstable.

Polar Plot

The same data as above is plotted on the complex plane as polar coordinates.

Bode Plot

The gain and phase are plotted, as separate curves, against frequency. The gain is usually expressed in decibels (dB) where

$$\text{gain in dB} = 20\ \log_{10}\ |\,\text{output/input}\,| \tag{3.58}$$

Nichols Plot

Here the gain and phase are plotted as one curve with increasing frequency marked. This method allows straightforward conversion of system data from open-loop to closed-loop forms, and vice versa.

Fuller discussions on all these methods are presented in D'Azzo and Houpis [25]; Van de Vegte [112]; Kuo [71].

These methods can be applied to the digital case by plotting the frequency response of the open-loop transfer function $G(z)$ when $z = e^{j\omega T}$. As ω increases from 0 to $\omega_s/2$, z moves round the unit circle, generating the frequency response of $G(z)$. Note that as ω is further increased, z repeatedly goes around the unit circle generating the same frequency response. Thus the frequency response of $G(z)$ over the interval $\omega T = [-\pi, \pi]$ is repeated over intervals $[\pm\pi, \pm3\pi]$, $[\pm3\pi, \pm5\pi]$, etc., owing to aliasing and the folding of higher frequencies onto the interval $[0, \omega_s/2]$, as already discussed in chapter 1.

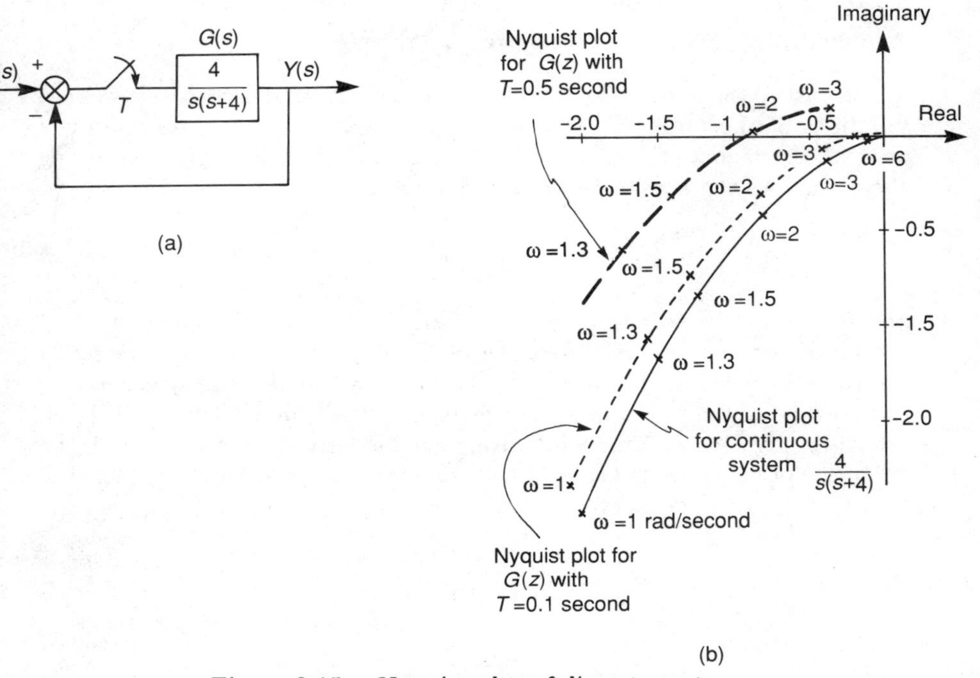

Figure 3.15 Nyquist plot of discrete systems

As an example, the servo system shown in Figure 3.15 is considered and its Nyquist plot determined. The open-loop transfer function is

$$G(s) = \frac{4}{s(s+4)} \tag{3.59}$$

which may be z-transformed, using tables, to

$$G(z) = \frac{z(1-e^{-4T})}{(z-1)(z-e^{-4T})} \tag{3.60}$$

To determine the frequency response we need to substitute

$$z = \exp(j\omega T) = \cos\omega T + j\sin\omega T \tag{3.61}$$

and evaluate the real and imaginary parts of $G(z = e^{j\omega T})$ as ω varies from 0 to $\omega_s/2$. The Nyquist plots for $T = 0.1$ s and 0.5 s are shown in Figure 3.15(b), together with the plot for the purely continuous system. It is clear that the discretisation destabilises the system as expected, and as the sampling rate is increased, we approach the continuous case. In addition, when deriving the data for even simple examples such as this, the complex nature of the expressions that have to be dealt with is evident; thus it is advisable to have the aid of computer programs for determining the frequency responses.

Approximate Frequency Domain Analysis

In contrast to the above "exact" frequency response analysis where awkward functions of the form $e^{j\omega T}$ have to be handled, an approximate form of analysis is adequate for most cases. This is made possible by transforming the z-plane to the w-plane, introduced in section 3.3, where

$$z = \frac{1+w}{1-w} \tag{3.62}$$

and $w = \sigma_w + jv$ defines a complex plane similar to the s-plane. To find the approximate frequency response of $G(w)$ we set $w = jv$ ($s = j\omega$ in the continuous case). To appreciate the w-domain analysis fully it is useful to be aware of the relationship between the "true" frequency ω and its w-domain equivalent, v. This relationship can be derived by starting with the definition of $w = (z-1)/(z+1)$ and comparing the respective frequencies, that is

$$\begin{aligned} [w]_{w=jv} &= \left[\frac{z-1}{z+1}\right]_{z=e^{j\omega T}} \\ &= \frac{\cos\omega T - 1 + j\sin\omega T}{\cos\omega T + 1 + j\sin\omega T} \end{aligned} \tag{3.63}$$

From straightforward analysis, equation (3.63) can be reduced to

$$v = \tan\frac{\omega T}{2} \tag{3.64}$$

which gives the result that the frequency in the w-plane is distorted. To overcome this problem, a new w' transformation can be introduced where

$$w' = \frac{2}{T}w \tag{3.65}$$

or

$$w' = \frac{2}{T}\frac{z-1}{z+1} \tag{3.66}$$

which gives

$$z = \frac{1+\frac{T}{2}w'}{1-\frac{T}{2}w'}. \tag{3.67}$$

Note this approximation is equivalent to Tustin's rule discussed earlier. Since $w' = \sigma_{w'} + jv'$, it defines another complex plane and the relationship between v' and ω can be shown to be (see Houpis and Lamont [46]) given by

$$v' = \frac{2}{T}\tan\frac{\omega T}{2}. \tag{3.68}$$

Table 3.8 Comparison of frequencies in s-, w- and w'-planes

s-plane ω rad/s	w-plane v rad/s	w'-plane v' rad/s
0.1	0.005	0.1
1	0.05	1
2	0.1	2
5	0.26	5.1
10	0.55	10.9
20	1.56	31.1
30	14.1	282

A numerical comparison of ω, v and v' for $T = 0.1$ s (sampling frequency $= 10$ Hz or 62.8 rad/s) is shown in Table 3.8. The frequencies of interest are $0 \leq \omega \leq \omega_s/2$ and so the highest frequency we can represent is 31.4 rad/s. Hence for high sample rates and low frequencies, $v' \approx \omega$ and for $\omega \leq \omega_s/4$, the approximation is good. The approximation can be further improved by using pre-warping techniques, see for example, Franklin and Powell [35].

We can therefore conclude that for most control systems (which are essentially low-pass), the w'-plane looks just like the s-plane in terms of actual quantities such as gains and bandwidths, and so the design procedure can be the same as that for the continuous case. Note that the major difference is that as v' varies from 0 to ∞ , it gives, in terms of the real frequency, ω, the range $0 - \omega_s/2$. Putting $\omega = \omega_s/2 \rightarrow \omega_s$ gives the w' frequency response corresponding to $v' = 0 \rightarrow -\infty$, and as we have already seen, the v' positive response is repeated for

$$\omega = n\omega_s \rightarrow (n + 1/2)\,\omega_s \qquad \text{for } n = 0, 1, 2, 3, \ldots \tag{3.69}$$

and the v' negative response is repeated for

$$\omega = (n + 1/2)\,\omega_s \rightarrow (n + 1)\,\omega_s \qquad \text{for } n = 0, 1, 2, 3, \ldots \tag{3.70}$$

Having obtained $G(w')$ it is then straightforward to apply all the continuous frequency domain design and analysis methods. The Bode analysis in particular of $G(jv')$ is much simpler to plot than $[G(z)]_{z=\exp(j\omega T)}$, since all the usual methods of breakpoints, straight line approximations, etc. can be applied.

We illustrate this by considering the rigid-body satellite control system discussed earlier, shown in Figure 3.16(a). We have

$$G(s) = \frac{1 - e^{-sT}}{s^3} \tag{3.71}$$

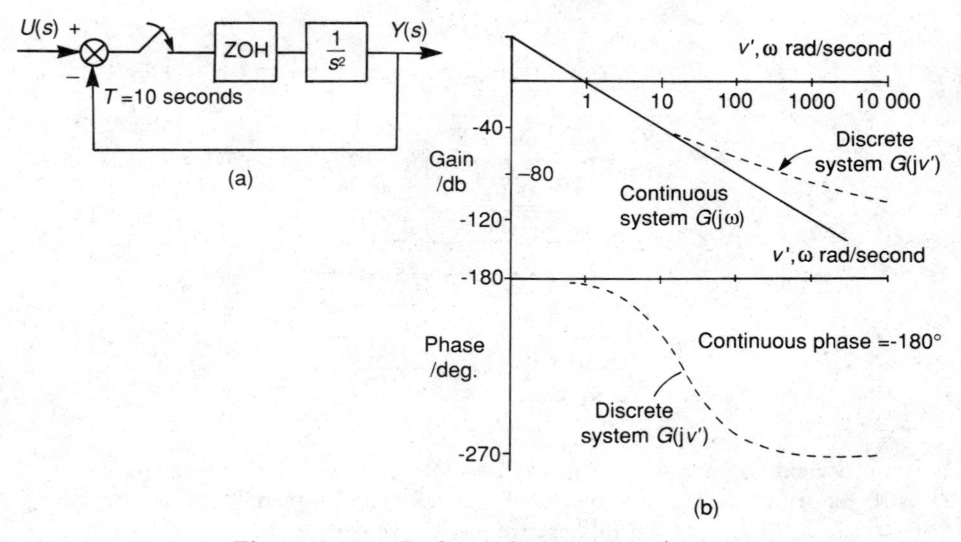

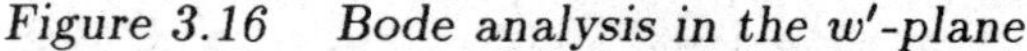

Figure 3.16 Bode analysis in the w'-plane

and so

$$G(z) = \frac{T^2}{2} \frac{z+1}{(z-1)^2} \tag{3.72}$$

Substituting

$$z = \frac{1 + \frac{T}{2}w'}{1 - \frac{T}{2}w'} \tag{3.73}$$

we get

$$G(w') = \frac{1 - w'/2T}{(w')^2} \tag{3.74}$$

To determine the frequency response in the w'-plane we let $w' = jv'$ and assuming $T = 10$ s we have

$$G(jv') = \frac{jv'/20 - 1}{(v')^2} \tag{3.75}$$

The Bode plot follows the usual rules and can be easily drawn, as shown in Figure 3.16(b). For comparison, the Bode response for the continuous system is also shown. Note that the main difference is the extra zero in the w' transfer function. Such zeros are added because of the sampling and holding operations in digital systems. Further discussion on this can be found in Kuo [70].

3.7 Summary

The main methods for analysing discrete control systems have been outlined. As already stated, the discussions are necessarily brief and as such cannot do justice to the amount of research that has been carried out to arrive at the results. For further reading the reader should consult the references cited for elaboration of the points mentioned. We now move on to controller design, where we will make use of the techniques introduced in this chapter.

3.8 Problems

1. Assess the stability of the servo system shown in Figure 3.17 by
   (i) using Jury's/Raible's method;
   (ii) using Routh's method;
   (iii) determining the closed-loop pole locations in the z-plane.

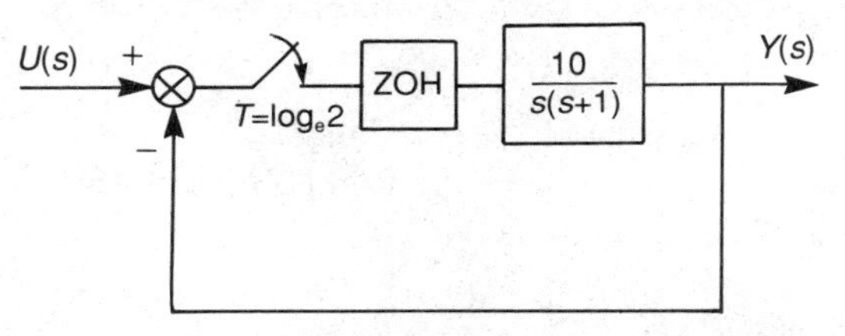

Figure 3.17 Stability assessment of servo system

2. For the systems whose characteristic equations are shown below, use Raible's method to establish how many poles are in the stable region, on the stability boundary and in the unstable region:
   (i) $z^3 - 5z^2 + 3z + 2 = 0$
   (ii) $z^3 - 1.5z^2 + 2z + 3 = 0$
   (iii) $z^4 - 0.9z^3 + 0.6z^2 + 0.4z - 0.3 = 0$
   (iv) $z^4 + 2.5z^3 + 2.1z^2 + 0.5z + 0.04 = 0$
   (v) $z^3 + 3.6z^2 + 3.88z + 2.4 = 0$
   (vi) $z^6 + 3.6z^5 + 4.05z^4 + 2.85z^3 + 2.15z^2 - 0.75z - 0.9 = 0.$
3. Determine the closed-loop discrete transfer function of the system shown in Figure 3.18(a). If

$$G_1(s) = 2\frac{\left(1 - e^{sT}\right)(s+5)}{s^2}, \quad G_2(s) = 25\frac{\left(1 - e^{sT}\right)}{(s+1)(s+25)},$$

   $H(s) = 1$, and $T = 0.5$ s, determine the time response of the system when the input is a unit step.
4. Figure 3.18(b) shows a sampled-data system. As K is varied
   (i) apply Routh's criterion to determine whether or not the system becomes unstable;

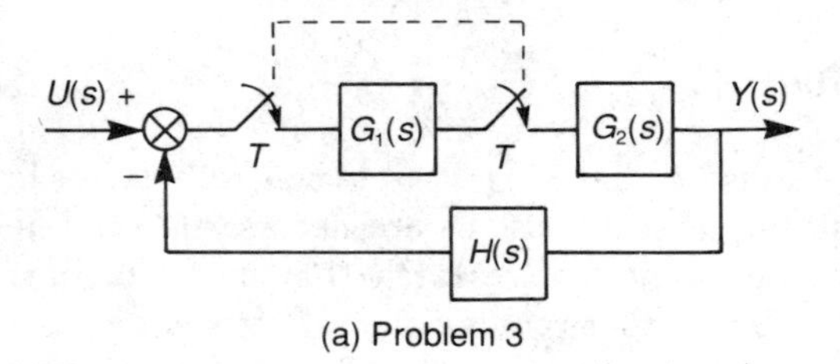

(a) Problem 3

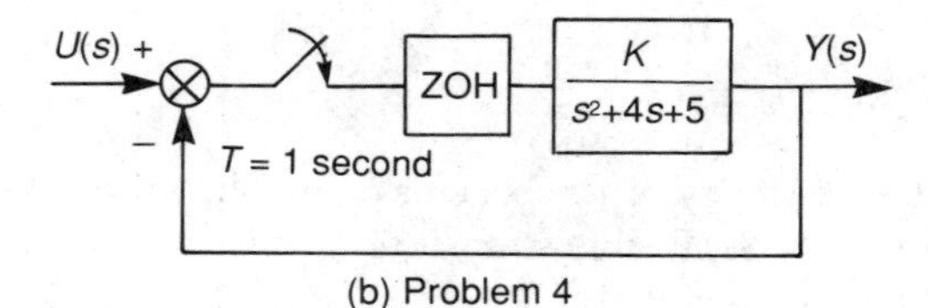

(b) Problem 4

Figure 3.18 Closed-loop sampled-data systems

(ii) sketch the root locus in the z-plane;
(iii) for $K = 10$, draw the Bode plot and establish the gain and phase margins of the system.

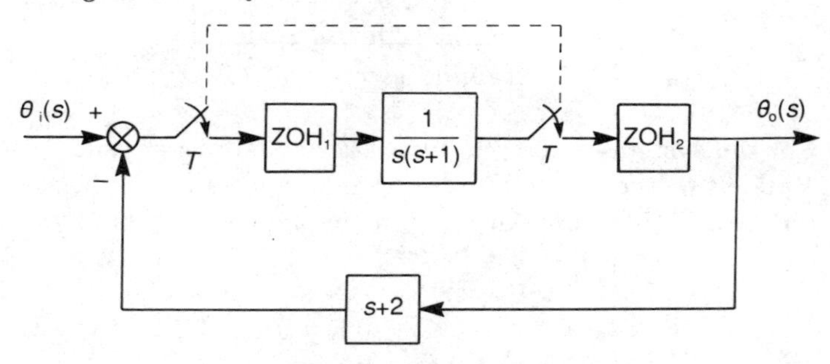

Figure 3.19 Compensated servo system

5. Figure 3.19 shows a compensated servo control system containing two synchronised samplers. If the period of the samplers is $T = \ln 2$ s, determine the output magnitude of ZOH_2 at the times $t = 0, T, 2T$ and ∞ when the input is a unit step.
6. The open-loop transfer function of a system is

$$G(s) = \frac{500K}{s^2 + 30s + 400}$$

This is used with a zero-order hold in a sampled-data feedback system whose sampling period $T = 0.1$ s.

Sketch the root locus of the characteristic equation for $0 \leq K < \infty$ in the z-plane. Calculate the value of K when the damping ratio is zero and find the damping ratio when $K = 1$. For $K = 1$ find the percentage overshoot of the closed-loop output response when the input is a unit step.

7. A digital control system, with $T = 0.1$ s has an open-loop transfer function of

$$G(z) = \frac{2K(z+0.5)}{(z-1)(z-0.5)}$$

Sketch its Nyquist plot for $\omega = 0 \rightarrow \omega_s/2$, and determine the critical value of K for closed-loop stability.

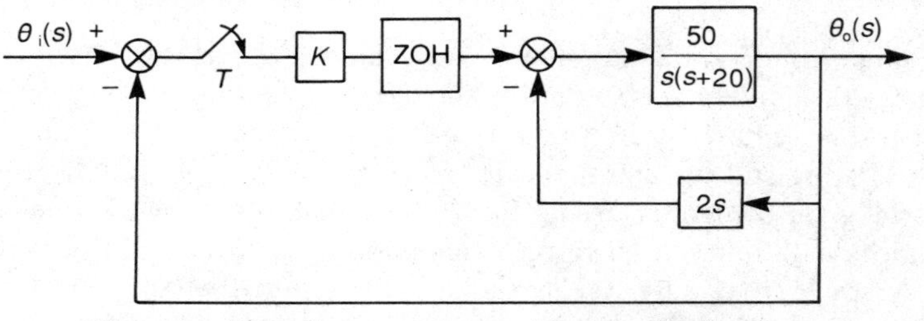

Figure 3.20 *Servo system with rate feedback*

8. Figure 3.20 shows a servo system with rate feedback being used in a closed-loop sampled data application. If the sampling rate is 50 Hz determine the range of K over which the system is stable.

4 Digital Compensator Design

4.1 Introduction

The most important component in any control loop is the compensator, the primary task of which is to modify the overall dynamics such that the system behaves as required. In this chapter we consider various methodologies for designing suitable compensators for digital control systems. A working knowledge of controller design for the continuous case is assumed although attempts will be made to refresh this knowledge as the digital methods are discussed. We begin by stating the objectives aimed at when performing the designs. In general the requirements are similar to those for the continuous case, and include the following:

1. good transient behaviour – this may or may not allow a degree of overshoot, but fast response is usually required;
2. good steady state behaviour, such that errors are minimised;
3. good disturbance rejection.

In addition, the controller algorithms designed have to be realisable, that is, it should be physically possible to perform all the calculation at any particular instant. This essentially means that, at time t_1, all necessary variables are measurable and known, and values of signals at future times $t > t_1$ are not required. For a digital system, a further requirement is evident, namely that the problems of aliasing need to be avoided, or at least minimised, in order to avoid the problems discussed in chapter 1.

The main design methods use feedback to form a closed-loop system, as this is more reliable and insensitive to parameter variations than the open-loop case. Cascade or feedback compensation as well as feedforward control can be used to modify the transfer function to give the required performance. For more details see for example Kuo [70], or Franklin and Powell [35]. In addition several design methods exist for deriving a suitable controller after the form of the compensator has been decided upon. We will discuss the common approaches by considering, in the main, the system shown in Figure 4.1. This can represent several practical applications, such as the attitude control of a rigid-body satellite or a magnetic suspension control system. For example, consider a satellite in orbit transmitting data down to the earth. For effective operation, the pointing angle of the satellite needs to be controlled accurately by firing thrusters in the X, Y and Z directions. To simplify the analysis we concentrate on just one plane. When a particular thruster is fired it causes the satellite to rotate about an axis. This relationship is given by Newton's Second Law of Motion in rotation form, which states that

$$\text{Torque}(t) = \text{moment of inertia } (I) \times \text{angular acceleration } (\ddot{y}(t)) \quad (4.1)$$

Assuming the moment of inertia to be unity gives

$$\text{Torque}(t) = \ddot{y}(t) \quad (4.2)$$

Laplace-transforming this and assuming zero initial conditions yields

$$\text{Torque}(s) = s^2 Y(s) \quad (4.3)$$

Hence the transfer function of the satellite moving in one angular plane is

$$G(s) = \frac{Y(s)}{\text{Torque}(s)} = \frac{1}{s^2} \quad (4.4)$$

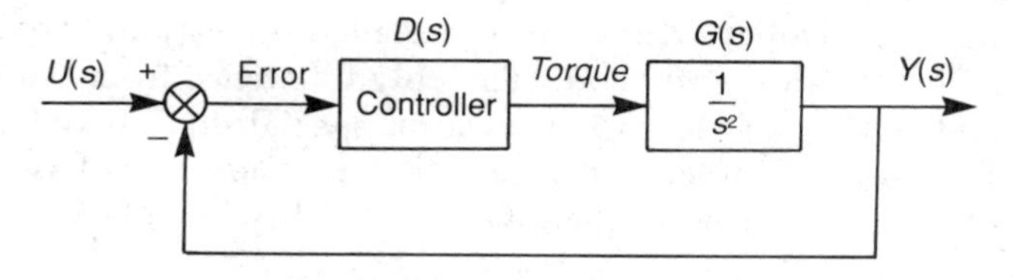

Figure 4.1 Control system under investigation

The continuous system can easily be shown to be marginally stable (giving undamped oscillations) without any compensator, because of the double integrator giving zero phase margin. The situation becomes worse when a sample and hold device is inserted into the control loop (see chapter 3), and a dynamic controller is required to stabilise the performance and bring it to within required specifications. We will perform this design using a number of approaches.

When designing control systems, a variety of criteria can be aimed for depending on the application and the precise objectives. In practice these reduce to "rules of thumb" such as:

- designing for a phase margin of $\approx 50°$,
- designing for a gain margin of ≈ 9 dB,
- designing for a damping ratio ζ of ≈ 0.7.

The precise rules can vary from application to application, but they are widely used in all areas of control engineering. In the following discussion, when the different design methods are illustrated, attempts will be made to use a uniform approach in the specifications, so that the different resulting controllers can be compared.

The design methods we shall consider are given below.

Continuous Domain Design
: A continuous controller $D(s)$ is designed for the analogue system $G(s)$ using frequency or time domain methods. $D(s)$ is then digitised to give $D(z)$. One must be conservative when performing the continuous design since the subsequent digitisation will introduce extra instability that must be allowed for.

Digital Design
: This involves starting with a system $G(z)$ so that digitisation effects have been taken into account. The transfer function is then transformed into the w' domain to give $G(w')$. Since the w' domain is similar to the frequency domain we can use the normal frequency methods to design a controller $D(w')$ which can then be transformed back into the z-domain to give $D(z)$.

Digital Root Locus Design
: A digital controller $D(z)$ can be designed directly in the z-domain using the root locus method. The only difference from the standard method is that the locus is drawn on special digital root locus paper as discussed in chapter 3, and that root locations have different interpretations, owing to the different stability boundary.

State Feedback Design
: The state-space approach can be applied to the difference equations. Controllability and observability criteria are as in the continuous case.

Digital PID Design
: This is the digital version of the classical three term controller, where the dynamic elements are approximated using numerical methods.

Deadbeat Response Design
: The controller in this case is designed so that the output settles exactly to the input in a minimal number of sampling instants. In a sense, the system is tuned to give an ideal response in a particular situation. For example when the input is a unit step at $t = 0$, the best output response we can hope for is zero at $t = 0$ and 1 for $t = T, 2T, 3T, \ldots$. Hence there is no overshoot and the only difference is due to the dead time introduced in the discretisation procedure. However, deadbeat designs are tuned to be ideal for a particular type of input and may give poor performance in other circumstances.

Optimal Control Design
: Control laws can be designed using a cost function that is optimised over time. This must be chosen carefully as all other aspects will be ignored in the optimisation procedure.

Controller Design in the Presence of Noise
: The controller design methods considered thus far assume noise-free conditions – in practice control systems are subject to many stochastic effects that can disturb the system. Steps can be taken so that the disturbances are minimised or catered for by other methods.

Having outlined the methods commonly used we now present some actual compensator designs for the system in Figure 4.1; specifications will be introduced as the designs proceed. When performing these designs it is useful to have access to CAD packages for analysing the control systems in the various domains. The main ones used by the author for performing the designs in this chapter are:

CODAS (Control System Design and Simulation) for PC – see the reference under Codas Operating Manual [20].

PC-MATLAB with Control Systems Toolbox – see the reference under PC-MATLAB User Guide [93].

4.2 Continuous Domain Design

We will design a continuous controller $D(s)$ for $G(s) = 1/s^2$, and then digitise to give $D(z)$. As mentioned earlier, the uncompensated system gives

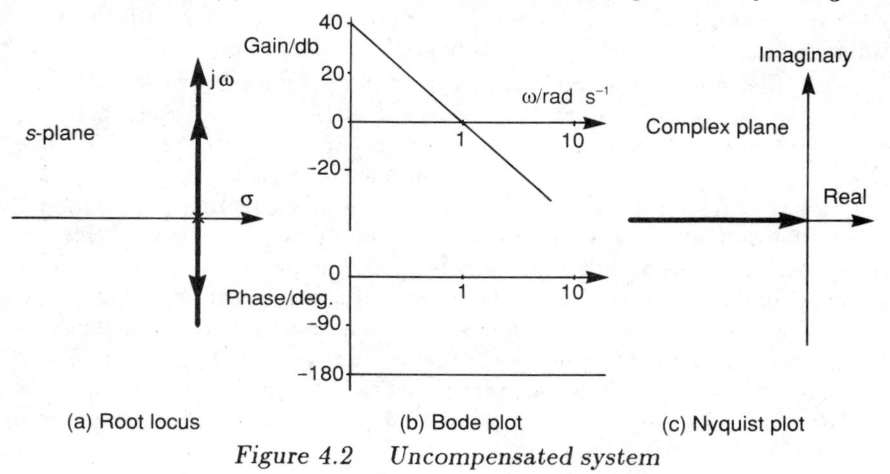

Figure 4.2 Uncompensated system

continuous undamped oscillations, as can be seen from the root locus (see Figure 4.2(a)), starting at the origin and going to infinity on the imaginary axes. Clearly the locus needs to be pulled over into the left half plane and Figures 4.2(b) and (c) show that the phase needs to be advanced. Assume that the closed-loop system specifications are:

$$\text{Damping ratio } \zeta \geq 0.5 \tag{4.5}$$

$$\text{Settling time } t_s \leq 1 \text{ second to a step input} \tag{4.6}$$

The uncompensated closed-loop transfer function (CLTF) is

$$\frac{Y(s)}{U(s)} = \frac{1}{s^2+1} \tag{4.7}$$

indicating zero damping and a natural undamped frequency $\omega_n = 1$ rad/s. A proportional plus derivative controller can give the necessary corrective action. However, a lead network of this type gives noise amplification problems at high frequencies and the gain needs to be limited. This is accomplished by the addition of a lag term whose breakpoint is at some (higher) frequency. Hence the controller required takes the form

$$D(s) = K\frac{(s+z_1)}{(s+p_1)} \tag{4.8}$$

The main problems yet to be solved are where to put the pole and zero, and the choice of the overall gain. The open-loop transfer function (OLTF) becomes

$$K\frac{(s+z_1)}{s^2(s+p_1)} \tag{4.9}$$

with the zero nearer the origin on the negative real axis. We will use (continuous domain) root locus analysis to perform the design.

From the standard second-order curves in appendix B, we see that the settling time t_s is approximately given by $4/\zeta\omega_n$ (for $\pm 2\%$ tolerance). Hence the undamped natural frequency needs to be approximately 8 rad/s. The closed-loop poles need to lie beyond the damping ratio line for $\zeta = 0.5$ at least this distance from the origin. Since zeros increase the amount of oscillations in any system, we need to design for a damping ratio of $0.6 - 0.9$ so that the specification will be met. The precise margin necessary is dependent on the locations of the poles and zeros. As an initial attempt, let us design for a ζ of approximately 0.8. The representation of poles on this line is $\alpha(-1 \pm j0.8)$ where $\alpha > 0$ is a real number. For $\alpha = 6.25$, the poles indicate an undamped natural frequency of 8 rad/s. However, to allow a margin of safety, we let $\alpha = 8$ which indicates that the closed-loop pole locations are at $-8 \pm j6.4$.

We need therefore to assign the controller pole and zero so that the root locus passes through, or near, these positions. A zero near the origin will cause the two poles to branch and arc towards the left half plane in a roughly circular path until they meet again on the negative real axis. Using straight forward linear algebra, a zero at -6.6 will cause roughly the required effect. The pole further along the negative axis will destabilise the system but if it is far enough along it will not have a significant effect over the region of interest. A factor of 10 is usually sufficient and so we position the pole to be at -66 giving the resulting OLTF as

$$K\frac{(s+6.6)}{s^2(s+66)} \tag{4.10}$$

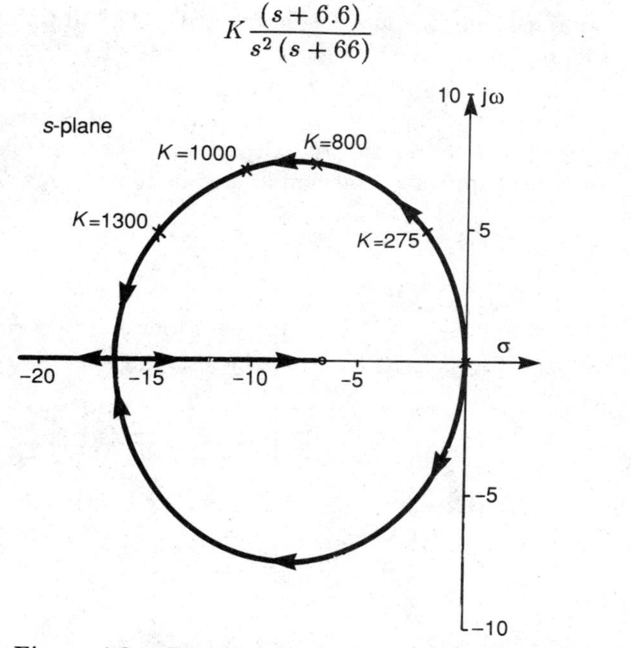

Figure 4.3 *Root locus of compensated system*

Using a CAD package, the root locus for this system is shown in Figure 4.3 from which we can ascertain that the required roots for $\zeta \approx 0.8$ imply a gain of ≈ 1000. If a suitable computer package is not available the task of obtaining the design value of K is much more difficult. The characteristic equation has to be studied and factorised for various values of K to establish the correspondence of sections of the root locus to ranges of K. This is extremely tedious when the calculation are made by hand, especially when high order polynomials are being considered. The author does not recommend this approach, and suggests that every effort be made to obtain access to a suitable CACSD (Computer Aided Control System Design) package.

The continuous controller $D(s)$, is therefore given by

$$D(s) = 1000\frac{(s+6.6)}{(s+66)} \tag{4.11}$$

and needs to be discretised in order to give the $D(z)$. Before carrying out the digitisation, a suitable sampling interval for the discrete system must be chosen so that the continuous system can be adequately represented and controlled. The damped frequency of the compensated system is approximately 6.4 rad/s which gives a frequency of approximately 1 Hz. From the sampling theorem, we must sample at least twice this rate in order to retain all the relevant information, but for good accuracy and minimal destabilising effects, we must sample much faster, typically 10 – 20 times faster. Let us use a sampling frequency of 20 Hz, that is $T = 0.05$ s.

With this value we can digitise $D(s)$ using either a numerical approximating procedure, or via the pole-zero mapping method (see chapter 2); for convenience we choose to use the latter method. Here the poles and zeros in the s-plane are mapped to poles and zeros in the z-plane. Hence $D(z)$ must have

(i) a zero at $z = e^{+sT} = e^{-6.6T} = 0.719$;
(ii) a pole at $z = e^{-66T} = 0.037$.

Therefore $D(z) = K_{DC}\frac{(z-0.719)}{(z-0.037)}$, where K_{DC} is found by ensuring that $D(s)$ and $D(z)$ have similar gains at low frequencies, hence

$$\left[1000\frac{(s+6.6)}{(s+66)}\right]_{s=0} = \left[K_{DC}\frac{(z-0.719)}{(z-0.037)}\right]_{z=1} \tag{4.12}$$

giving $K_{DC} = 342.7$. The digital controller required is therefore

$$D(z) = 342.7\frac{(z-0.719)}{(z-0.037)} \tag{4.13}$$

Let us now evaluate the performance of the controller designed. In order to do this we must first digitise the system so that the z-transform techniques can be applied. In this digitisation, a zero-order hold is usually inserted so that the impulsive inputs from the digital controller are transformed into a piecewise constant signal (see Figure 4.4(a)). The open-loop discrete transfer function of the compensated system is $D(z)G(z)$, where

$$G(z) = \mathcal{Z}\left\{\frac{1-e^{-sT}}{s^3}\right\} = \frac{T^2}{2}\frac{(z+1)}{(z-1)^2} \tag{4.14}$$

Hence

$$D(z)G(z) = 0.428\frac{(z-0.719)(z+1)}{(z-1)^2(z-0.037)} \tag{4.15}$$

giving rise to a closed-loop discrete transfer function of

$$\frac{Y(z)}{U(z)} = \frac{0.428z^2 + 0.12z - 0.308}{z^3 - 1.609z^2 + 1.194z - 0.345} \tag{4.16}$$

The response of this to a unit step input is shown in Figure 4.4(b) where we can see that the system does not correspond to $\zeta = 0.5$ since this would give $\approx 17\%$ overshoot whereas in practice we have over 70% overshoot.

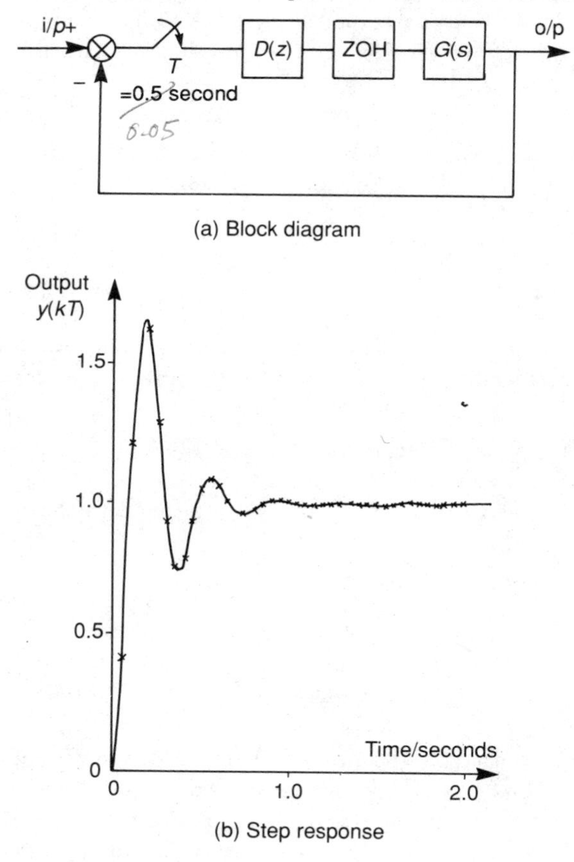

Figure 4.4 Compensated system (continuous design)

Clearly the approximation is not adequate in this case. Increasing the sampling rate will improve the situation but the main reasons for the discrepancy cannot be avoided by using the continuous design methodology. This is because even if $D(z)$ generates exactly the same sample values as $D(s)$, the zero-order hold reconstruction is only an approximation to the continuous input to $G(s)$ assumed in the design of $D(s)$. Other sources of error are also present since

(i) At each sampling instant the computer measures the error and needs to process it before the control signal is ready for implementation. A finite delay time is therefore introduced in the control loop which, as is well known, causes extra phase lag that destablises the system. For a better design, this delay needs to be considered and allowed for in the controller.

(ii) A second-order system is assumed so that the analysis is much simpler, but this is contrary to the fact that we are dealing with a third-order plant.

The design can be improved by repeating the procedure with a finer degree of approximation and a bigger safety margin. We do not intend to do this but turn instead to our next method since the continuous design approach has been demonstrated adequately.

4.3 Digital Design

In this section we will repeat the above design exercise but will now use the discrete system transfer function as the starting point. From section 4.2 we have

$$G(z) = \mathcal{Z}\{ZOH + G(z)\} = 0.00125\frac{(z+1)}{(z-1)^2} \tag{4.17}$$

for a sampling interval $T = 0.05$ s. Transforming this into the w'-plane by letting

$$z = \frac{1+\frac{T}{2}w'}{1-\frac{T}{2}w'} \tag{4.18}$$

we get

$$G(w') = \frac{1-0.025w'}{(w')^2} \tag{4.19}$$

Note the addition of the non-minimum phase zero due to the sampling and holding operation. Extra zeros are usually introduced when analysing discrete systems in the w'-domain and it is necessary to ensure that they are included in the design procedure. Now, as shown in chapter 3, the w'-plane is similar to the s-plane and therefore we can use continuous design procedures to determine $D(w')$. We will use frequency domain analysis in the w'-plane to do this by letting $w' = jv'$ (compare $s = j\omega$). Hence

$$G(jv') = \frac{1-j0.025v'}{(jv')^2} \tag{4.20}$$

The Nyquist and Bode plots of this are shown in Figure 4.5, where it is clear that compensation is required to stabilise the performance and bring it to within the specifications. At the 0 dB crossover $v' = 1$ rad/s and we have zero phase margin. From standard second-order curves (see appendix B) it can be seen that we require about 50° phase margin for a damping ratio $\zeta \approx 0.5$. Although this is easily achieved by the use of a phase-lead network,

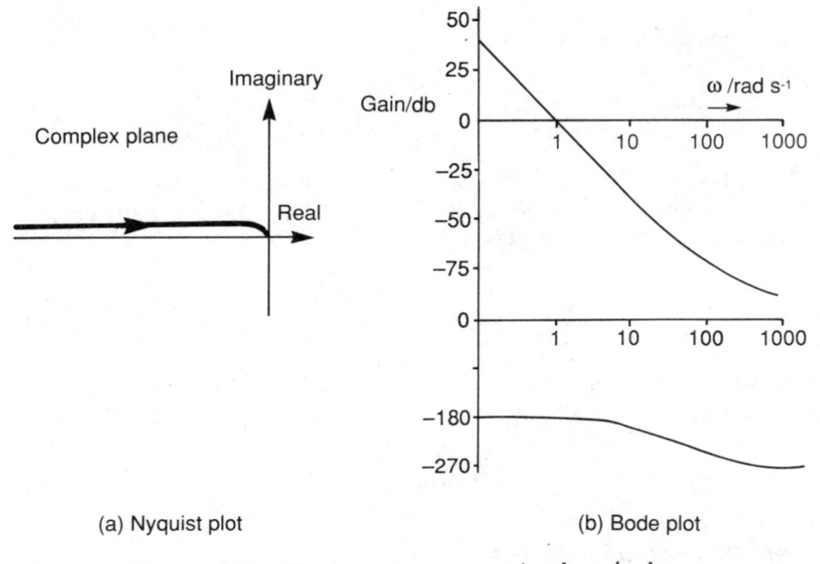

(a) Nyquist plot (b) Bode plot

Figure 4.5 Frequency response in the w'-plane

another important problem also needs to be tackled. The bandwidth of the system is rather low and hence the settling time specifications will not be satisfied. To increase the 0 dB crossover, we need to insert gain into the open-loop transfer function, in order to speed up the system. However, in doing so, extra phase advance will be necessary to bring the performance to within the required constraints. A compromise needs to be made between how far to the right we move the 0 dB crossover point and the amount of phase advance necessary to achieve this. Unfortunately there are no rigid rules for dealing with conflicts of this kind. The control engineer must iterate around the design and validation procedure until a suitable controller is obtained.

As a first attempt we will start by designing a 0 dB crossover at 10 rad/s, that is by inserting a gain of 100 in the open-loop transfer function. We then need to insert a phase advance controller to give the required phase margin. This has the following form

$$D_1(w') = \frac{p_1}{z_1}\frac{(w' + z_1)}{(w' + p_1)} \tag{4.21}$$

The system has a phase of $-194°$ at 10 rad/s. If $z_1 = 2$, a phase advance of 78.7° is introduced at 10 rad/s. Hence the compensator pole can add 15° phase lag. If $p_1 = 60$, a phase lag of approximately 9.5° is added. The proposed controller is therefore

$$D_1(w') = 30\frac{(w'+2)}{(w'+60)} \tag{4.22}$$

which causes the 0 dB point to be changed since at $v' = 10$ rad/s, a gain of approximately 5 is introduced by the controller. Therefore to keep the 0 dB crossing at 10 rad/s we require

$$D_1(w') = 6\frac{(w'+2)}{(w'+60)} \tag{4.23}$$

Taking the DC gain into account the overall controller is calculated from $D(w') = K_{DC}D_1(w')$ so that

$$D(w') = 600\frac{(w'+2)}{(w'+60)} \tag{4.24}$$

Inverse w'-transforming by letting

$$w' = \frac{2}{T}\frac{(z-1)}{(z+1)} \tag{4.25}$$

we get (for $T = 0.05$ s)

$$D(z) = 252\frac{(z-0.9)}{(z+0.2)} \tag{4.26}$$

To evaluate the performance of this design we need to study the following discrete open-loop transfer function:

$$D(z)G(z) = 0.315\frac{(z+1)(z-0.9)}{(z-1)^2(z+0.2)} \tag{4.27}$$

The closed-loop step response of this, shown in Figure 4.6, is reasonably close to the required specification and we can terminate the design with the above $D(z)$. Obviously if the performance is unsatisfactory we can iterate round the design loop once more, changing the damping ratio and/or the bandwidth of the system accordingly.

Comparing the continuous design method of section 4.2 with this digital design we can see the advantages of the latter. Although the w'-domain approach requires a little more effort, the accuracy one can achieve by using this design procedure makes it well worth while. The main reason for the improvement over the digitised continuous design is the fact that the sample and hold approximations are allowed for during the design stage, since the starting point is the discrete transfer function $G(z)$.

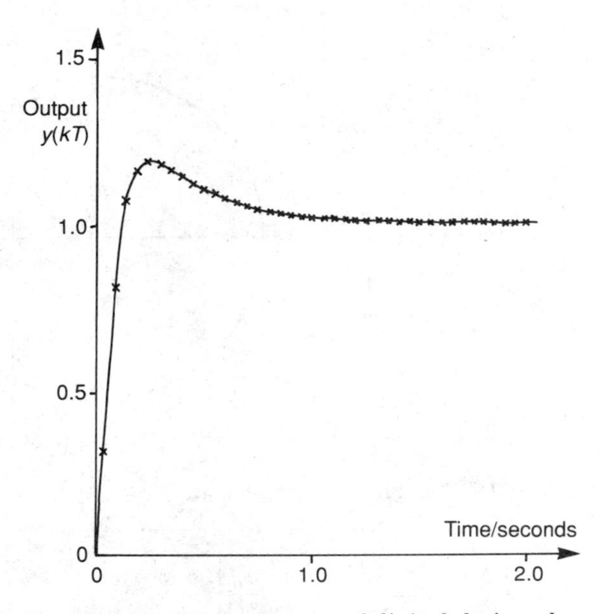

Figure 4.6 Step response of digital designed system

4.4 Digital Root Locus Design

The root locus of a system is a plot showing the variation of the closed-loop poles while some parameter (usually gain) is varied from zero to infinity. This can be used to design a suitable controller since root locations define the characteristics of the system under study. The rules for drawing the digital version of the locus are exactly as in the continuous domain, but the locations have different interpretations as discussed in chapter 3.

The uncompensated system is shown in Figure 4.7(a), which has a characteristic equation of $1 + KG(z) = 0$. The root locus of this is shown in Figure 4.7(b), where we can see that the system is unstable for all gains, and as before, some form of dynamic compensator is required. We shall attempt to modify the locus so that the required specifications are met, that is $\zeta \geq 0.5$ and $t_s \leq 1$ s. This may be achieved by inserting poles and zeros into the open-loop transfer function. The question is: "At what locations?" A zero needs to be put on the positive axis to pull the locus into the stable region so that it intersects the required damping curve. To deduce this we need to look at where we want the root locus to lie. As shown in chapter 3, to satisfy the settling time requirements (for $\approx 2\%$ error tolerance) the closed-loop poles must lie within a circle of radius $e^{(-4T/t_s)}$. Letting $T = 0.05$ s, as before means a radius of ≈ 0.8. Hence, allowing for a degree of safety, let us design a compensator such that the root locus passes through

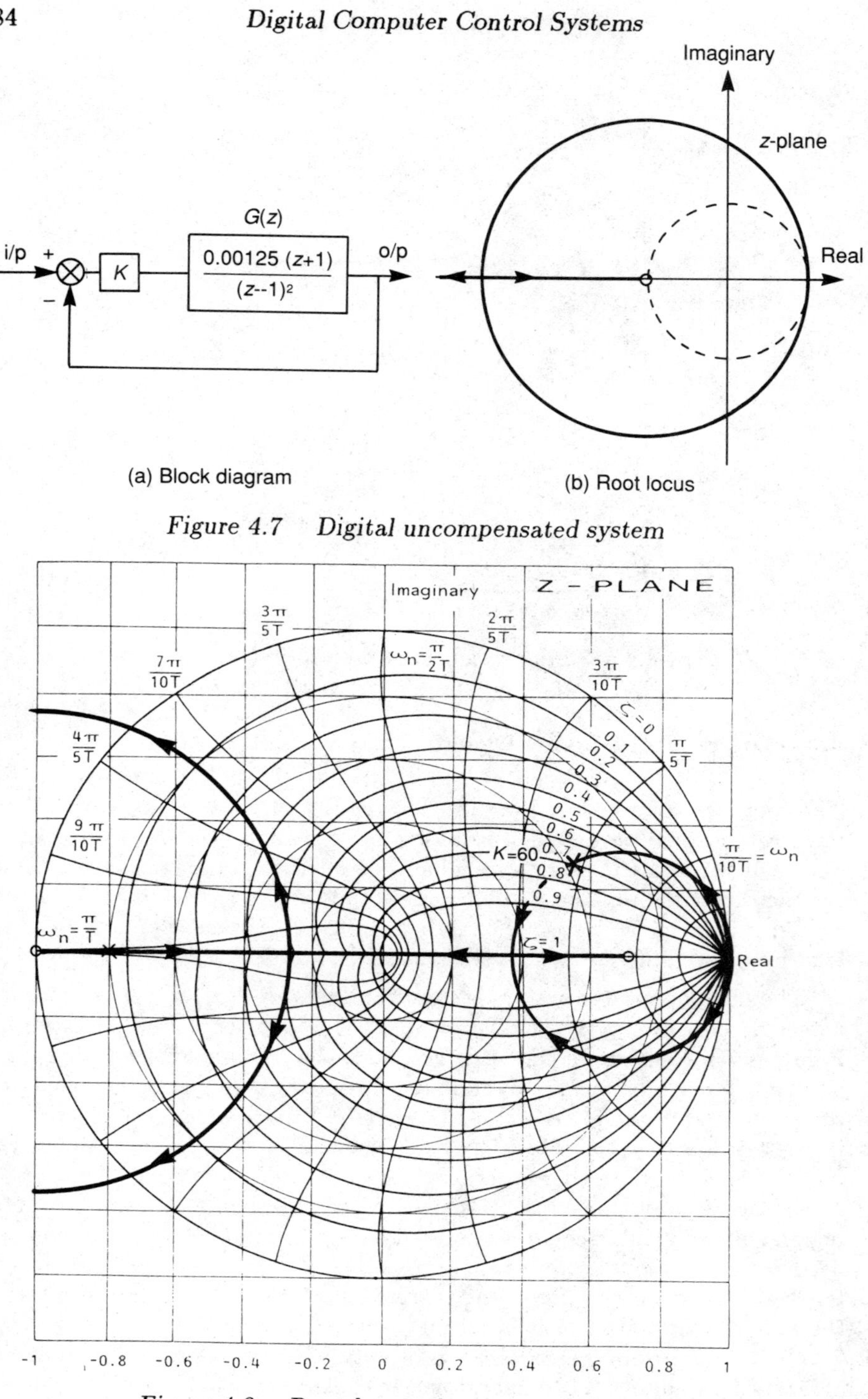

Figure 4.7 **Digital uncompensated system**

Figure 4.8 **Root locus of compensated system**

the intercept between the circle of radius 0.7 and the $\zeta = 0.6$ locus, that is approximately through the point $0.65 \pm j0.3$. A zero placed near $z = 1$ will cause the two poles to become complex and follow an anti-clockwise circular path unless influenced by other terms. A zero at $z = 0.7$ will cause the locus to pass near the required point. The pole can be added on the negative axis within the required radius for settling time requirements, but far enough away so as not to affect the dominant part of the locus. Let us put the pole at $z = -0.8$. Hence the proposed controller is

$$D(z) = K\frac{(z - 0.7)}{(z + 0.8)} \tag{4.28}$$

The root locus of the system with this compensator is shown in Figure 4.8.

The gain required to give the above closed-loop poles is 600. As with all designs, the resulting controller needs to be assessed and we do this by

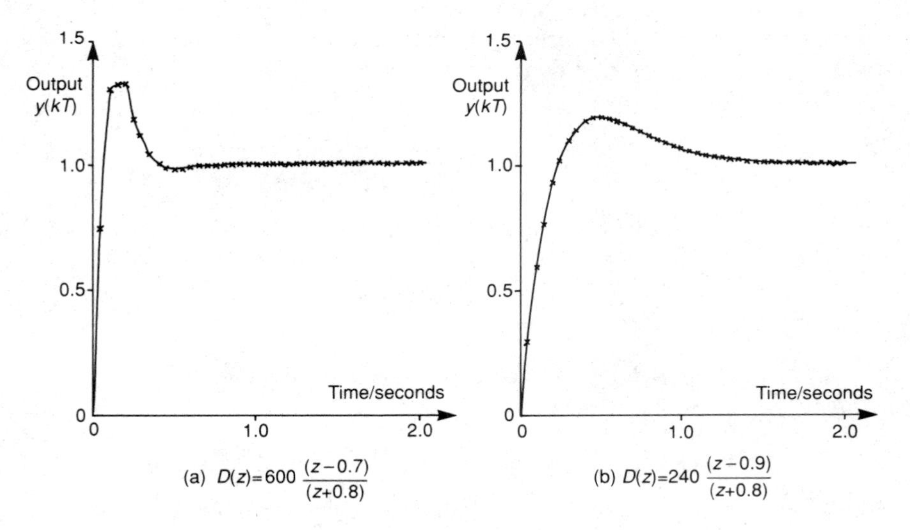

Figure 4.9 Step responses for root locus designs

performing a step response test; Figure 4.9 shows the result of this test, where we can see an overshoot of almost 30% although the settling time is well within specification (about 0.5 s). We can therefore repeat the design by using a slower settling time which will require less gain and consequently less overshoot. Repeating the design procedure, the controller

$$D(z) = 240\frac{(z - 0.9)}{(z + 0.8)} \tag{4.29}$$

can be calculated, which gives a step response as shown in Figure 4.9(b). Here we can see less overshoot but the settling time has been increased. We can obviously improve on this if we wish to speed the system further, but we stop here and proceed to our next design method.

4.5 State Feedback Design

In this approach a state-space is used to represent the behaviour of the plant under consideration. The state-space is a vector space of dimension equal to the order of the system, and is useful for studying the internal "states" as well as input/output relations. The digital transfer function is transformed into a state-space form by the use of classical analogue programming techniques. We illustrate this by considering the above system where

$$G(z) = \frac{Y(z)}{U(z)} = 0.00125\frac{(z+1)}{(z-1)^2} \tag{4.30}$$

There are several approaches for obtaining a state-space form; we will demonstrate the so-called direct programming method. Here we start by dividing the numerator and denominator by the highest power in z, namely by z^2. This gives

$$\frac{Y(z)}{U(z)} = 0.00125\frac{(z^{-1}+z^{-2})}{1-2z^{-1}+z^{-2}} \tag{4.31}$$

which can be split into two equations by defining a dummy variable $E(z)$ such that

$$\frac{Y(z)}{E(z)} = (z^{-1}+z^{-2}); \qquad \frac{E(z)}{U(z)} = \frac{0.00125}{1-2z^{-1}+z^{-2}} \tag{4.32}$$

Noting that z^{-1} is a delay of 1 sampling interval we have

$$Y(z) = z^{-1}E(z) + z^{-2}E(z) \tag{4.33}$$

$$E(z) = 0.00125U(z) + 2z^{-1}E(z) - z^{-2}E(z) \tag{4.34}$$

Assuming $E(z)$ is available we can construct the block diagram shown in Figure 4.10. The states are now defined to be the outputs of the z^{-1} blocks. We therefore have

$$zX_1(z) = X_2(z) \tag{4.35}$$

$$zX_2(z) = -X_1(z) + 2X_2(z) + 0.00125U(z) \tag{4.36}$$

$$Y(z) = X_1(z) + X_2(z) \tag{4.37}$$

Converting to the time domain and writing in matrix form we have the required state-space as:

$$x((k+1)T) = \begin{bmatrix} 0 & 1 \\ -1 & 2 \end{bmatrix} x(kT) + \begin{bmatrix} 0 \\ 0.00125 \end{bmatrix} u(kT) \tag{4.38}$$

$$y(kT) = \begin{bmatrix} 1 & 1 \end{bmatrix} x(kT) \tag{4.39}$$

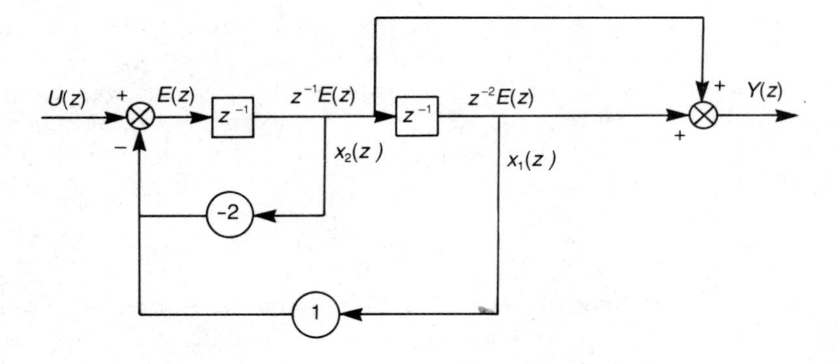

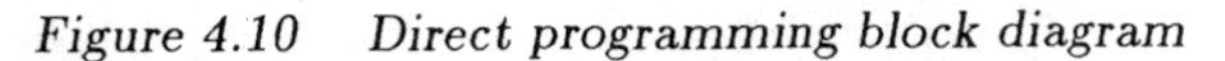
Figure 4.10 Direct programming block diagram

or in compact form

$$x((k+1)T) = Ax(kT) + Bu(kT) \tag{4.40}$$

$$y(kT) = Cx(kT) \tag{4.41}$$

where for an n-dimensional system with m inputs and p outputs we have

$$\begin{array}{lcll} A & = & n \times n & \text{system matrix} \\ B & = & n \times m & \text{input matrix} \\ C & = & p \times n & \text{output matrix} \end{array} \tag{4.42}$$

and x, u and y are vectors with $x \in R^n$, $u \in R^m$ and $y \in R^p$. Figure 4.11 shows this matrix form in relation to the input/output transfer function.

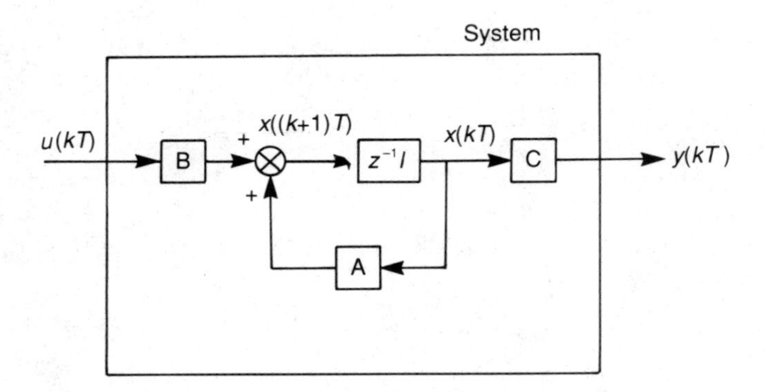

Figure 4.11 Block diagram of state-space form

As in the continuous case we can use eigenvalues, eigenvectors, similarity transformations, etc. (see for example Chen [19]). It can easily be shown that converting from a transfer function approach to a state-space representation gives a suitable minimal realisation. However when transforming from a state-space form to the transfer function form we may find terms cancelling (pole-zero cancellation). This is due to the phenomena called controllability and observability, for which there are the following definitions:

Controllability

A system is controllable if there exists a realisable control sequence such that we can transfer the system state from any initial state to any final state in a finite time. It can be shown that a system is controllable if the matrix

$$\left[B,\ AB,\ A^2B,\ \cdots,\ A^{n-1}B\,\right] \tag{4.43}$$

has full rank n.

Observability

A system is observable if it is possible to determine the state vector by using only input and output values. This is true if the matrix

$$\begin{bmatrix} C \\ CA \\ CA^2 \\ \vdots \\ CA^{n-1} \end{bmatrix} \tag{4.44}$$

has full rank n.

Controllability and observability give rise to four possible subsystems in any system, namely observable (S_o), observable and controllable (S_{co}), controllable (S_c) and uncontrollable and unobservable (S_u). These subsystems are shown in Figure 4.12, where the links that are present or missing in the overall system structure can be clearly seen. S_{co} is the subsystem that relates to the transfer function and forms the basis of the classical design and analysis methods. It can easily be shown that the transfer function can be calculated using

$$\frac{Y(z)}{U(z)} = C\left[zI_n - A\right]^{-1}B \tag{4.45}$$

Having briefly introduced state-space concepts we can discuss compensation within this framework. The A matrix mentioned above determines the eigenvalues (or poles) of the system which defines the dynamic behaviour. The classical (feedback) compensation techniques can be extended to the state-space by the introduction of control loops that generate the input by

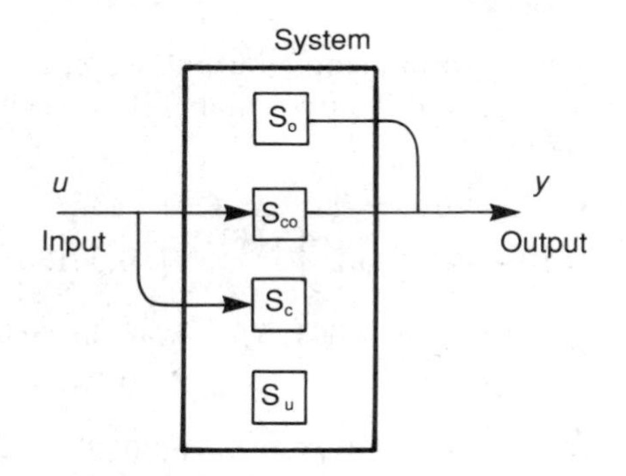

Figure 4.12 Controllable and observable subsystems

a linear combination of the state x, as is shown in Figure 4.13.

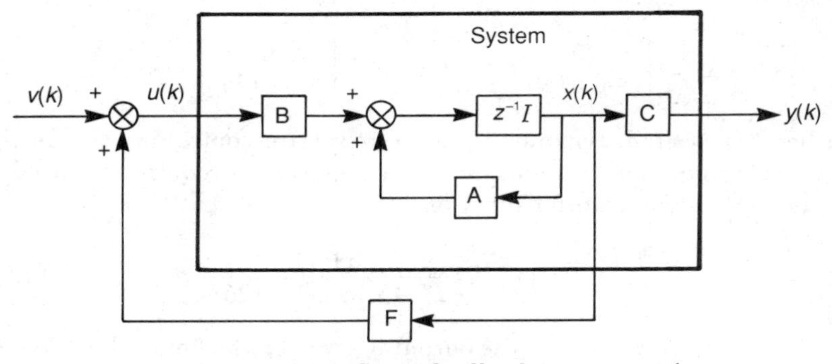

Figure 4.13 State feedback compensation

Since $u(k) = v(k) + Fx(k)$, the state-space representation becomes

$$x(k+1) = (A+BF)\,x(k) + Bv(k) \tag{4.46}$$

Hence the response of the system is no longer determined by A but by the matrix $(A+BF)$. Now it turns out that if the system is controllable, we can design a feedback matrix so that the closed-loop poles are at any desired locations, see Kuo [70] or Van de Vegte [112].

Our system above is controllable (and observable) – hence we can assign the poles at arbitrary locations. From chapter 3 we see that the closer the poles are to the origin in the z-plane, the quicker the dynamic behaviour. As in section 4.4 for a sampling interval of 0.05 s we require the poles to lie within a radius of 0.8 and on the 0.5 damping ratio line. Allowing for a degree of safety let us design a closed-loop system with poles at $0.65 \pm j0.3$. Hence the characteristic equation we require is

$$z^2 - 1.3z + 0.5125 = 0 \tag{4.47}$$

For our second-order single-input single-output system let the feedback matrix $F = [f_1, f_2]$, where f_1 and f_2 are scalars. Hence with $u = v + Fx$ the state equation becomes

$$x(k+1) = \begin{bmatrix} 0 & 1 \\ f_1' - 1 & f_2' + 2 \end{bmatrix} x(k) + \begin{bmatrix} 0 \\ 0.00125 \end{bmatrix} v(k) \tag{4.48}$$

where $f_1' = 0.00125 f_1$ and $f_2' = 0.00125 f_2$. The characteristic equation of equation (4.48) is

$$z^2 - (2 + 0.00125 f_2)\, z + 1 - 0.00125 f_1 = 0 \tag{4.49}$$

Comparing this with the desired characteristic equation, we can deduce that $f_1 = 390$ and $f_2 = -560$. Hence the required feedback vector is $F = [390, \ \ -560]$. The $-$ve sign indicates negative feedback and the closed-loop state equation is

$$x(k+1) = \begin{bmatrix} 0 & 1 \\ -0.5125 & 1.3 \end{bmatrix} x(k) + \begin{bmatrix} 0 \\ 0.00125 \end{bmatrix} v(k) \tag{4.50}$$

which has the desired dynamics. However, we are controlling the output $y(k)$ of the system and so to ensure that it behaves as required we need to study the closed-loop transfer function

$$\frac{Y(z)}{V(z)} = 0.00125 \frac{(z+1)}{z^2 - 1.3z + 0.5125} \tag{4.51}$$

Hence the steady state value of the output is given by the final value theorem as

$$y_{ss} = \lim_{z \to 1} \ \frac{z-1}{z} \ \frac{0.00125\,(z+1)}{z^2 - 1.3z + 0.5125} V(z) \tag{4.52}$$

When $V(z) = z/(z-1)$ we have $y_{ss} = 0.0011765$ and so a large steady state error exists. The reason for this is that we have used state feedback merely to modify the dynamics of the system so that the transient behaviour is as required. No account is taken of whether the system responds to the input in any specified way. Hence we have considered the so-called "regulator problem". This is where the external input is zero and the system has been designed to bring any initial error to zero within specified transient constraints. When the system output is required to follow command signals, we have the situation known as "the servo control problem", which we now discuss.

State Feedback in Servo Systems

In servo systems a general requirement is that we have at least one integrator in the control loop so that steady state errors will be eliminated. One method of introducing an integrator is to introduce a new state vector that integrates the difference between the command signal $r(k)$ and the system output $y(k)$. Figure 4.14 shows a possible configuration for designing servo systems with state feedback and integral control action.

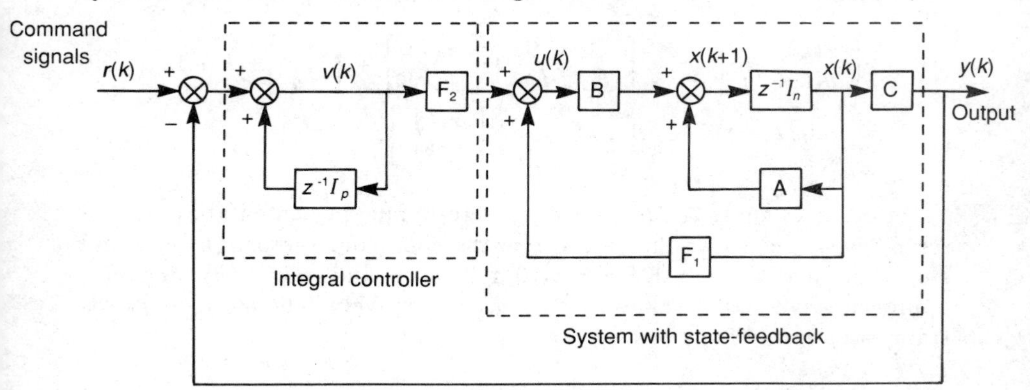

Figure 4.14 Servo system with state feedback and integral control

The integral controller consists of p integrating elements, one for each command input component, and the gain matrices $F_1(m \times n)$ and $F_2(m \times p)$ need to be designed for a suitable solution. The system equations are

$$x(k+1) = Ax(k) + B(k) \tag{4.53}$$

$$y(k) = C(k) \tag{4.54}$$

$$v(k) = v(k-1) + r(k) - y(k) \tag{4.55}$$

where $v(k)$ is the p-dimensional error vector. Equation (4.55) can be written as

$$\begin{aligned} v(k+1) &= v(k) + r(k+1) - y(k+1) \\ &= v(k) + r(k+1) - C\,(Ax(k) + Bu(k)) \\ \text{hence } v(k+1) &= -CAx(k) + v(k) - CBu(k) + r(k+1) \end{aligned} \tag{4.56}$$

The control $u(k)$ is given by

$$u(k) = F_1x(k) + F_2v(k) \tag{4.57}$$

and so

$$\begin{aligned} u(k+1) &= F_1x(k+1) + F_2v(k+1) \\ &= (-F_1 + F_1A - F_2CA)x(k) \\ &\quad + (I_m + F_1B - F_2CB)\,u(k) + F_2r(k+1) \\ \text{hence } u(k+1) &= K_1x(k) + K_2u(k) + F_2r(k+1) \end{aligned} \tag{4.58}$$

where $K_1 = -F_1 + F_1A - F_2CA$, and $K_2 = I_m + F_1B - F_2CB$. Noting that $u(k)$ is a linear combination of the state vectors $x(k)$ and $v(k)$, we can define a new state vector consisting of $x(k)$ and $u(k)$ to get a new state-space description as

$$\begin{bmatrix} x(k+1) \\ u(k+1) \end{bmatrix} = \begin{bmatrix} A & B \\ K_1 & K_2 \end{bmatrix} \begin{bmatrix} x(k) \\ u(k) \end{bmatrix} + \begin{bmatrix} 0 \\ F_2 \end{bmatrix} r(k+1) \quad (4.59)$$

$$y(k+1) = \begin{bmatrix} C & 0 \end{bmatrix} \begin{bmatrix} x(k) \\ u(k) \end{bmatrix} \quad (4.60)$$

In order to apply the pole-placement technique presented above to this servo system, consider the case where the command vector $r(k)$ is a unit step such that $r(k) = 1$ for $k = 0, 1, 2, 3, \ldots$, and $x(k), u(k)$, and $v(k)$ approach steady state values x_{ss}, u_{ss}, v_{ss} respectively. Therefore, at steady state, equation (4.55) gives

$$v_{ss} = v_{ss} + 1 - y_{ss} \quad (4.61)$$

and so $y_{ss} = 1$, and there is no steady state error in the output. Let us define the error vectors by

$$x_e(k) = x(k) - x_{ss} \quad (4.62)$$

$$u_e(k) = u(k) - u_{ss} \quad (4.63)$$

which can be used to describe the system equations as

$$\begin{bmatrix} x_e(k+1) \\ u_e(k+1) \end{bmatrix} = \begin{bmatrix} A & B \\ F_{11} & F_{12} \end{bmatrix} \begin{bmatrix} x_e(k) \\ u_e(k) \end{bmatrix} \quad (4.64)$$

where $F_{11} = -F_1 + F_1A - F_2CA$ and $F_{12} = I_m + F_1B - F_2CB$. It is straightforward to show that this system is controllable if the system defined by equation (4.53) is controllable (see Ogata [89]). The dynamics of the closed-loop servo system are determined by the eigenvalues of the system matrix in equation (4.64). For our above example we have

$$-F_1 + F_1A - F_2CA = [-f_{11} - f_{12} + f_2, \qquad f_{11} + f_{12} - 3f_2] \quad (4.65)$$

and

$$I_m + F_1B - F_2CB = 1 + 0.00125(f_{12} - f_2) \quad (4.66)$$

The closed-loop system matrix is therefore

$$\begin{bmatrix} 0 & 1 & 0 \\ -1 & 2 & 0.00125 \\ -f_{11} - f_{12} + f_2 & f_{11} + f_{12} - 3f_2 & 1 + 0.00125(f_{12} - f_2) \end{bmatrix} \quad (4.67)$$

which has a characteristic equation of

$$\begin{aligned} z^3 &- (3 - 0.00125(-f_{12} + f_2))z^2 + (3 + 0.00125(-f_{11} + f_{12} + f_2))\, z \\ &- 1 + 0.00125 f_{11} = 0 \end{aligned} \quad (4.68)$$

Assume, as above, that our closed-loop poles are at $0.65 \pm j0.3$ and 0.5, in order to satisfy the settling time requirements. Hence the desired characteristic equation is

$$z^3 - 1.8z^2 + 1.1625z - 0.25625 = 0 \quad (4.69)$$

Comparing the coefficients of equations (4.68) and (4.69) we have

$$3 - 0.00125(-f_{12} + f_2) = 1.8 \quad (4.70)$$

$$3 + 0.00125(-f_{11} + f_{12} + f_2) = 1.1625 \quad (4.71)$$

$$-1 + 0.00125 f_{11} = -0.25625 \quad (4.72)$$

From (4.72) we have $f_{11} = 595$ which we can use in equations (4.70) and (4.71) to give $f_{12} = -917.5$, and $f_2 = 42.5$.

The output response of this servo design is shown in Figure 4.15, where we can see that it satisfies the required specifications.

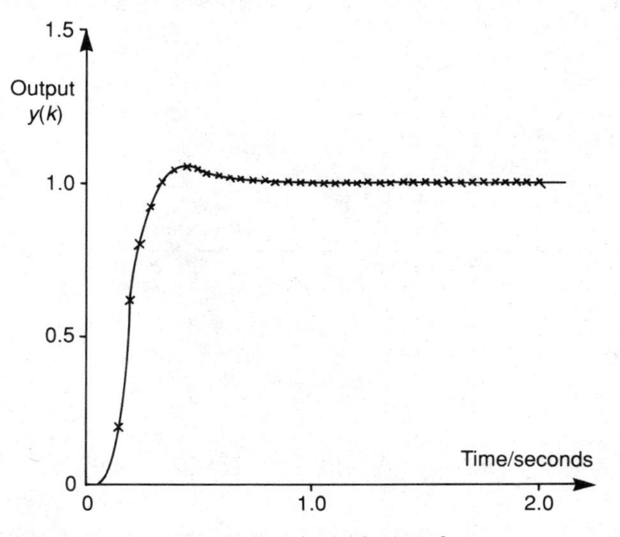

Figure 4.15 State feedback design for servo system

State Observers

It is worth pointing out at this stage that the state feedback design relies on the complete state vector being available for feedback purposes, which in practice is not the case. Measuring all the state variables is usually not possible, because of financial constraints. Also, in many cases, the states do not have physical meaning, and so it is impossible to measure them. Fortunately, an easier solution exists to get over this general unavailability of the state vector where a state estimator (or state observer) can be constructed using knowledge of the A, B, and C system matrices (see for example Chen [19]). Using the principle of separation, this state estimate can be used as if it were the real state vector, in the application of state feedback laws and other state variable designs.

A state observer can be used to estimate the state vector, x by $\hat{x}$, as shown in Figure 4.16 (assuming that the system is observable).

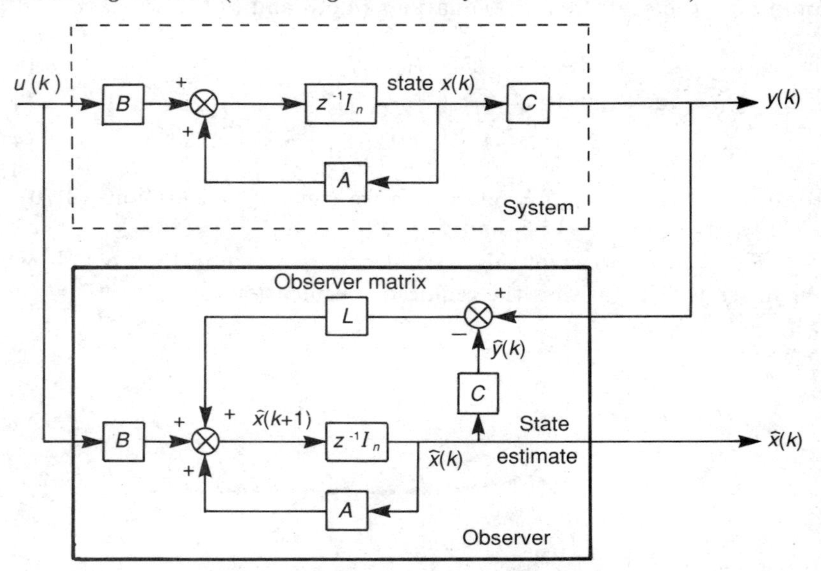

Figure 4.16 Full-order state observer

The dynamical equation of the observer is seen to be

$$\hat{x}(k+1) = (A - LC)\,\hat{x}(k) + Ly(k) + Bu(k) \tag{4.73}$$

Letting $e = x - \hat{x}$ be the error between the system and estimated state vectors, it is readily seen that

$$e(k+1) = (A - LC)\,e(k) \tag{4.74}$$

By suitable selection of the observer matrix L, it is possible to drive this error to zero arbitrarily fast. It is normal practice to design the observer poles, that is, the poles of $(A - LC)$ to be approximately $2 - 3$ times faster than the (compensated) system poles. In the z-domain this means the observer poles are closer to the origin. Having obtained an estimate of the state vector, it turns out that, because of the separation property, $\hat{x}$ can be used in place of x without affecting the state feedback designed results, see Chen [19].

It is possible to construct a full-order observer as discussed above, but in most cases some of the states are directly available at the system output, and so only those that are not accessible need estimation. Such a reasoning leads to the determination of the reduced-order observer, the main points of which we now outline.

Consider the system

$$x(k+1) = Ax(k) + Bu(k) \tag{4.75}$$
$$y(k) = Cx(k) \tag{4.76}$$

where $x \in R^n, u \in R^m$, and $y \in R^p$ with $n > p$ in general. To determine which states are directly available, and which need estimation we form an $n \times n$ matrix R such that $R = \begin{bmatrix} C \\ C_1 \end{bmatrix}$. The $(n-p) \times n$ matrix C_1 is arbitrarily chosen as long as det $[R] \neq 0$. Performing a similarity transformation using $\bar{x} = Rx$ gives an alternative state-space representation

$$\bar{x}(k+1) = RAR^{-1}\bar{x}(k) + RBu(k) \tag{4.77}$$
$$y(k) = CR^{-1}\bar{x}(k) \tag{4.78}$$

Partitioning this new state-space $\bar{x}$ into $\begin{bmatrix} \bar{x}_1 \\ \bar{x}_2 \end{bmatrix}$, where $\bar{x}_1$ are the first p elements of $\bar{x}$, and $\bar{x}_2$ are the last $(n-p)$ elements of $\bar{x}$ gives the system equations as

$$\begin{bmatrix} \bar{x}_1(k+1) \\ \bar{x}_2(k+1) \end{bmatrix} = \begin{bmatrix} \bar{A}_{11} & \bar{A}_{12} \\ \bar{A}_{21} & \bar{A}_{22} \end{bmatrix} \begin{bmatrix} \bar{x}_1(k) \\ \bar{x}_2(k) \end{bmatrix} + \begin{bmatrix} \bar{B}_1 \\ \bar{B}_2 \end{bmatrix} u(k) \tag{4.79}$$
$$y(k) = \begin{bmatrix} I_p & 0 \end{bmatrix} \begin{bmatrix} \bar{x}_1(k) \\ \bar{x}_2(k) \end{bmatrix} \tag{4.80}$$

Hence $\bar{x}_1$ are the state variables that are directly available at the system output, and $\bar{x}_2$ are the ones that need to be estimated. Following similar reasoning to that of the full-order observer, the dynamical equation for the estimate of $\bar{x}_2$, that is $\hat{\bar{x}}_2$, is found to be

$$\hat{\bar{x}}_2(k+1) = \left(\bar{A}_{22} - \bar{L}\bar{A}_{12}\right)\hat{\bar{x}}_2(k) + \bar{L}\left\{y(k+1) - \bar{A}_{11}y(k) - \bar{B}_1u(k)\right\} + \left(\bar{A}_{21}y(k) + \bar{B}_2u(k)\right) \tag{4.81}$$

This shows that at time k, future values of the outputs y at time $(k+1)$ need to be available for evaluating the right-hand side of the equation for $\hat{\bar{x}}_2$. We can eliminate this requirement by defining

$$q(k) = \hat{\bar{x}}_2(k) - \bar{L}y(k) \tag{4.82}$$

which gives

$$q(k+1) = \left(\bar{A}_{22} - \bar{L}\bar{A}_{12}\right) q(k) + \left(\bar{A}_{22} - \bar{L}\bar{A}_{12}\right) \bar{L}y(k) + \left(\bar{A}_{21} - \bar{L}\bar{A}_{11}\right) y(k) + \left(\bar{B}_2 - \bar{L}\bar{B}_1\right) u(k) \quad (4.83)$$

and therefore $q + \bar{L}y$ is an estimate of $\bar{x}_2$. Defining the error dynamics by $e = \bar{x}_2 - \hat{\bar{x}}_2$, it is straightforward to show that

$$e(k+1) = \left(\bar{A}_{22} - \bar{L}\bar{A}_{12}\right) e(k) \quad (4.84)$$

The error can be made to decrease to zero arbitrarily fast since $\left(\bar{A}_{22}, \bar{A}_{12}\right)$ is observable if the original system is observable, see Chen [19]. The complete state estimate is

$$\hat{\bar{x}} = \begin{bmatrix} \hat{\bar{x}}_1 \\ \hat{\bar{x}}_2 \end{bmatrix} = \begin{bmatrix} y \\ \bar{L}y + q \end{bmatrix} \quad (4.85)$$

Also $\bar{x} = Rx$ or $x = R^{-1}\bar{x} = Q\bar{x} = [Q_1 \quad Q_2] \begin{bmatrix} \bar{x}_1 \\ \bar{x}_2 \end{bmatrix}$, or

$$\hat{x} = Q\hat{\bar{x}} = \begin{bmatrix} Q_1 & Q_2 \end{bmatrix} \begin{bmatrix} I_p & 0 \\ \bar{L} & I_{n-p} \end{bmatrix} \begin{bmatrix} y \\ q \end{bmatrix} \quad (4.86)$$

which gives an estimate of the original state vector. The construction of the reduced-order observer is shown more clearly in Figure 4.17.

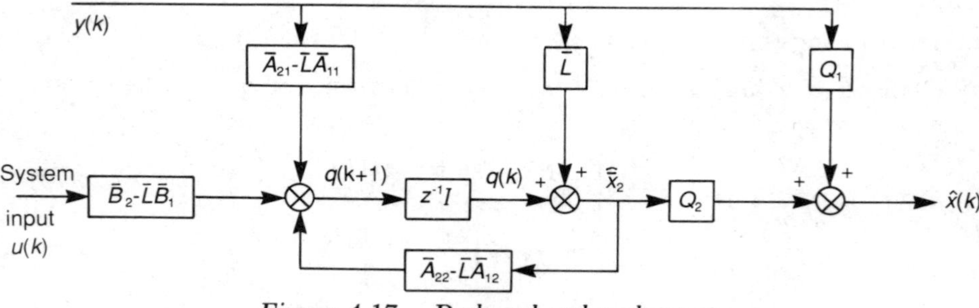

Figure 4.17 Reduced-order observer

The advantages of the state-space approach over the classical methods are that greater insight into the internal behaviour of the system is possible, as well as the ability to analyse individual sections of the overall system. The classical transfer function based techniques can only permit the design and analysis of the complete input/output description.

4.6 Digital PID Control Design

In the process industries the most widely used controller is the so-called "three-term" (or PID) controller. This, as the name suggests, has three components which are made up of a proportional term, an integral term and a derivative term. In the continuous case the PID controller has the form

$$D(s) = K_P + \frac{K_I}{s} + K_D s \tag{4.87}$$

where the three parameters K_P, K_I and K_D need to be chosen to give the desired behaviour. An alternative way of representing the PID controller is

$$K\left(1 + \frac{1}{T_I s} + T_D s\right) \tag{4.88}$$

where the parameters K, T_I, T_D define the three term controller. The two forms are equivalent and should cause no confusion. We prefer to use the form as expressed in equation (4.87) since it expresses the "gains" associated with each term more clearly. The PID controller is still extensively used in practice, and so a digital version is obviously necessary for computer control applications where the continuous operations need to be discretised and approximated. Consider the controller at the error point in a closed-loop system as shown in Figure 4.18.

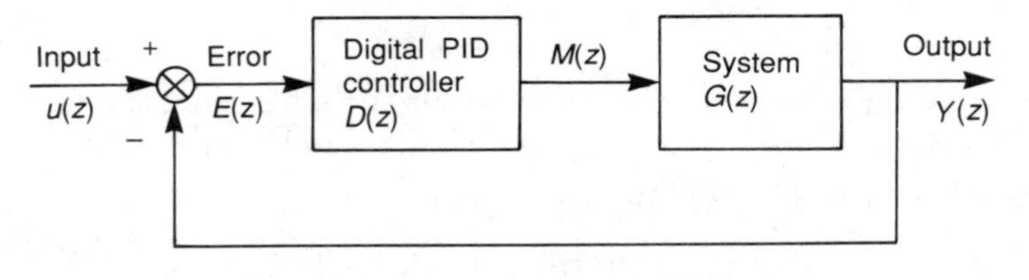

Figure 4.18 PID control

The output of the three term controller (in the time domain) is

$$m^*(kT) = K_P e^*(kT) + K_I i^*(kT) + K_D d^*(kT) \tag{4.89}$$

where i^* and d^* are the integral and derivative of the error signal respectively. The integral term is proportional to the area under the error signal as shown in Figure 4.19(a), and can be approximated using a number of methods, some of which are as follows.

Euler method

Here only the current error value is used so that

$$i^*(kT) = i^*((k-1)T) + Te^*(kT) \tag{4.90}$$

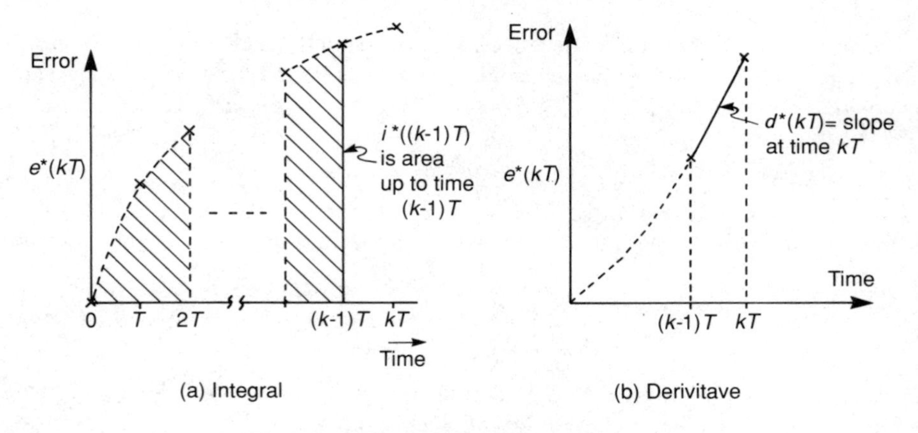

Figure 4.19 Digital approximation

Trapezium rule

Here the current and last error values are used to give

$$i^* (kT) = i^* ((k-1)T) + \frac{T}{2} (e^* (kT) + e^* ((k-1)T)) \tag{4.91}$$

Other higher order approximations are also possible using the well known Runge-Kutta methods, see for example Lapidus and Seinfeld [72].

The derivative of e^* is easily calculated using

$$d^* (kT) = \frac{e^* (kT) - e^* ((k-1)T)}{T} \tag{4.92}$$

These approximations can be studied in the z-domain by using the fact that z^{-1} is a delay of T seconds. Hence

$$\frac{I(z)}{E(z)} = \frac{Tz}{z-1} \tag{4.93}$$

for the Euler method, and

$$\frac{I(z)}{E(z)} = \frac{T}{2} \frac{(z+1)}{(z-1)} \tag{4.94}$$

for the Trapezium rule approximation. Similarly, we can show

$$\frac{D(z)}{E(z)} = \frac{z-1}{zT} \tag{4.95}$$

for the derivative term.

In most applications the system under consideration is unknown and a PID controller is used as a matter of course. To obtain adequate control performance the three gains need to be adjusted using plant trial data. Initial gain settings can be obtained using the classical methods of Ziegler and Nichols [123] which we now outline.

Closed-loop Methods

The first of these, called the continuous oscillation method, involves testing the closed-loop system using only proportional action. Starting with low gain values the plant is operated and the gain slowly increased until the plant output is oscillating at a constant amplitude. The plant plus proportional controller is now at the limit of stability. The period of oscillation T_u and the value of gain K_u are measured.

Then the settings suggested by Ziegler and Nichols [123] are as follows.

(i) For a proportional controller

$$K_P = 0.5K_u \tag{4.96}$$

(ii) For a proportional and integral controller

$$K_P = 0.45K_u, \quad K_I = \frac{1.2K_p}{T_u} = \frac{0.54K_u}{T_u} \tag{4.97}$$

(iii) For a proportional, integral and derivative controller

$$K_P = 0.6K_u, \quad K_I = \frac{1.2K_u}{T_u}, \quad K_D = \frac{0.6T_uK_u}{8} \tag{4.98}$$

The second method avoids the need for the sustained oscillations since these can be hazardous. In the damped oscillation method the gain is increased as above but only until the closed-loop decay ratio is at the desired value; a quarter decay ratio (see Coon [22], [23]) is commonly used. If the gain is K' at this point and the period of oscillation is T', the required PID control settings are

$$K_P = K', \quad K_I = \frac{1.5K'}{T'}, \quad K_D = \frac{T'K'}{6} \tag{4.99}$$

Open-loop Method

As the name implies, the feedback loop is broken and the response of the plant to a step input is recorded as shown in Figure 4.20. Assuming that this can be modelled by a simple gain, dead time and first-order lag (that is a first-order system), the Ziegler–Nichols settings are given below.

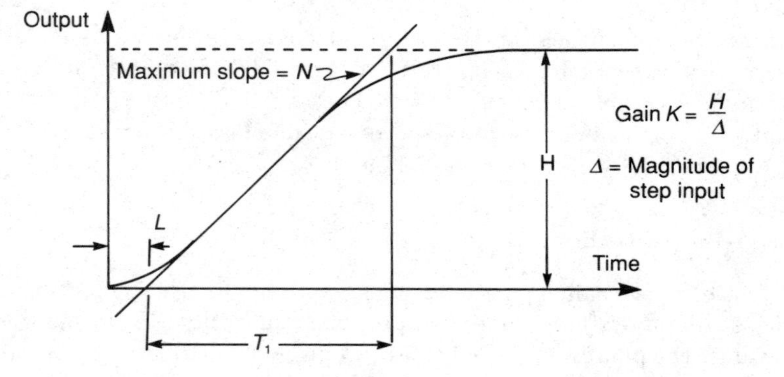

Figure 4.20 Open-loop system step response

(i) Proportional controller

$$K_P = \frac{\Delta}{NL} \tag{4.100}$$

(ii) Proportional and integral controller

$$K_P = \frac{0.9\Delta}{NL}, \qquad K_I = \frac{0.3K_P}{L} \tag{4.101}$$

(iii) Proportional, integral and derivative controller

$$K_P = \frac{1.2\Delta}{NL}, \qquad K_I = \frac{0.5K_P}{L}, \qquad K_D = 0.5LK_P \tag{4.102}$$

It is normal practice to use one of the above settings and make fine tuning adjustments to obtain the precise desired response. Other practical issues also need to be taken into account for the successful implementation of a PID controller – the main considerations are:

(i) actuator saturation; in the event of large commands a phenomenon known as integral windup can occur where the integral term builds up to a large value which causes actuator saturation and results in large overshoots and undershoots from the setpoint. It is necessary to impose upper and lower limits on the integrator so that such large terms are avoided; and

(ii) noise. In sampled-data systems, high frequency noise signals (present in most applications) may produce, when sampled, low frequency disturbances because of the folding of frequencies as discussed in chapter 1. These disturbances may have a significant effect on the performance of the control system and therefore it is necessary to insert a low-pass filter before sampling signals.

These and other considerations for PID controller are discussed more fully in Bennet [13] and Franklin *et al.* [36].

It should be noted that the closed-loop Ziegler–Nichols design method is only applicable to systems that are stable for low values of gain. Also the open-loop method works only for type 0 systems, where there are

- no poles at the origin in the s-plane,
- no poles at $z = 1$ in the discrete case,

so that K is finite. Unfortunately, neither of these methods is suitable for our double integrator system. We can still use a PID type of controller but the gain settings have to be arrived at by some other means. The integral term is not really required since the system has already two poles at $z = 1$. Assuming a P+D controller is used we have

$$D(z) = K_P + K_D \frac{z-1}{zT} \tag{4.103}$$

and using $T = 0.05$ s as before, the open-loop transfer function of the compensated system is

$$D(z)G(z) = 0.00125 \frac{(K_P + 20K_D)(z+1)\left(z - \frac{20K_D}{(K_P+20K_D)}\right)}{z(z-1)^2} \tag{4.104}$$

Suitable values for K_P and K_D can be obtained by using one of the design procedures already described. By root locus analysis we see that putting a zero at $z = 0.8$ will cause the double pole to migrate into the stable region. The root locus of this is shown in Figure 4.21(a) where we can see that the best response occurs at approximately $K_P + 20K_D = 210$. Solving this together with $\frac{20K_D}{(K_P+20K_D)} = 0.8$ as simultaneous equations gives $K_P = 42$, and $K_D = 8.4$ and the overall open-loop transfer function for this design (Design 1) is

$$D_1(z) = 0.2625 \frac{(z+1)(z-0.8)}{z(z-1)^2} \tag{4.105}$$

The closed-loop unity step response is shown in Figure 4.21(b), where we can see an overshoot of approximately 45%. As mentioned previously we can improve on this by reducing the gain and allowing for a longer settling time. Repeating the design with the zero placed at $z = 0.9$ gives rise to the root locus in Figure 4.22(a), and the required design gain is at approximately $K_P + 20K_D = 140$. Solving the equations for this situation gives $K_P = 14$ and $K_D = 6.3$, and the open-loop transfer function for this design (Design 2) is

$$D_2(z) = 0.175 \frac{(z+1)(z-0.9)}{z(z-1)^2} \tag{4.106}$$

The step response of this is shown in Figure 4.22(b) and can be seen to be an improvement over Design 1. The design of course can be improved further if required by performing another iteration.

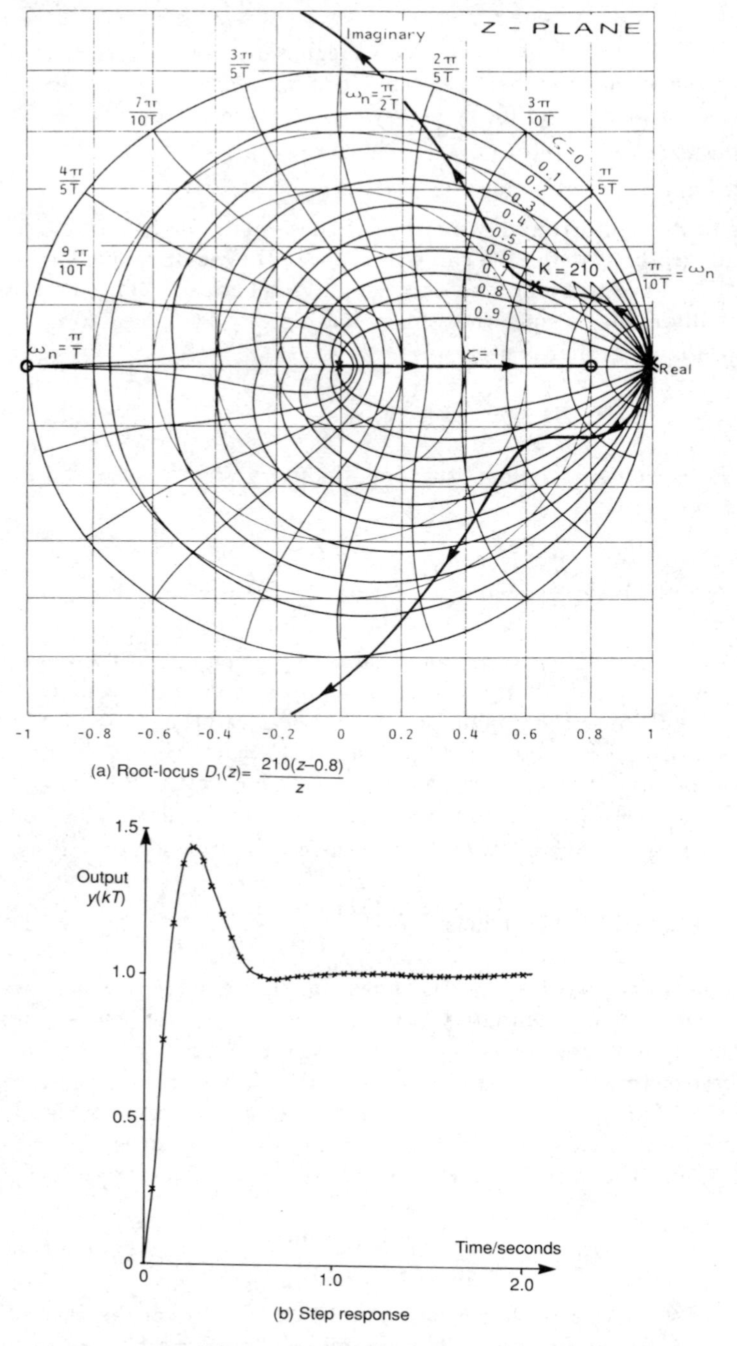

(a) Root-locus $D_1(z) = \frac{210(z-0.8)}{z}$

(b) Step response

Figure 4.21 P+D controller (Design 1)

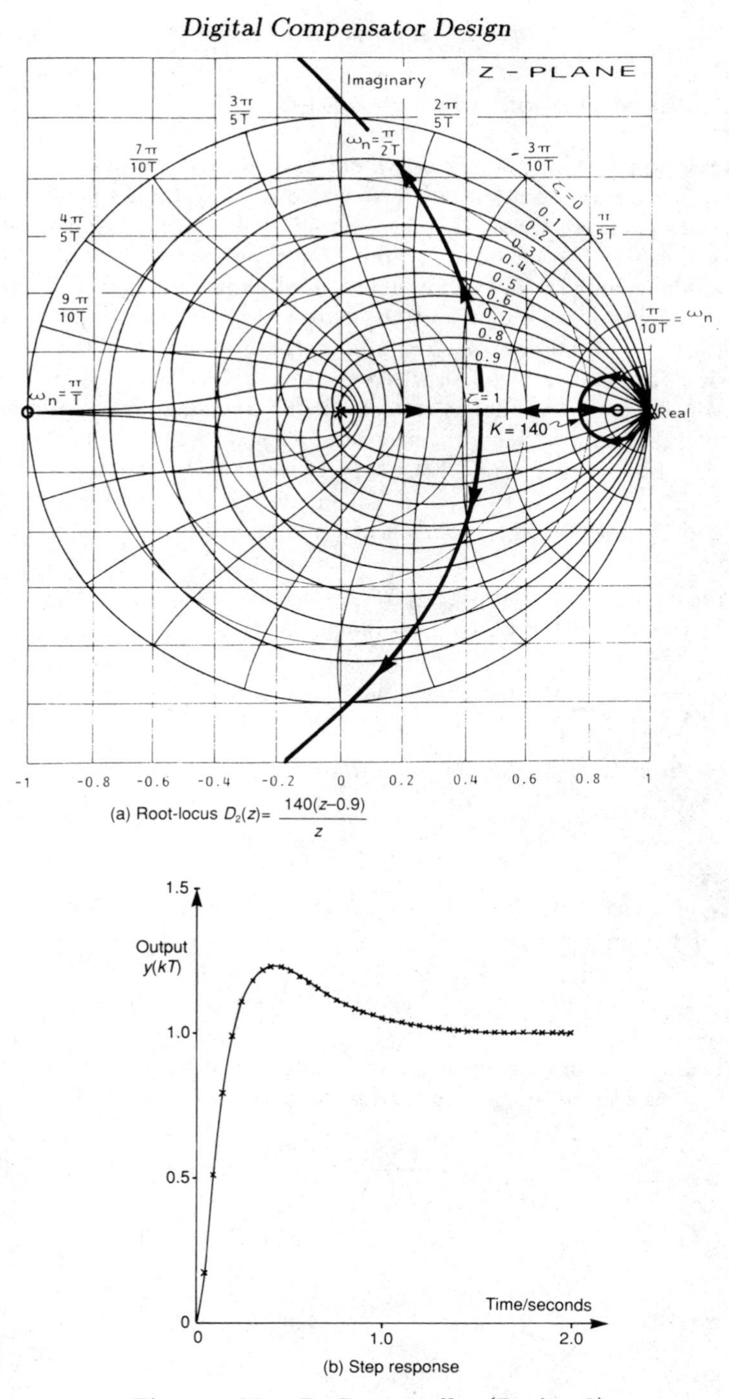

(a) Root-locus $D_2(z) = \dfrac{140(z-0.9)}{z}$

(b) Step response

Figure 4.22 P+D controller (Design 2)

4.7 Deadbeat Response Design

In contrast to the continuous case where all classical compensation design evaluations are rather subjective, in sampled data systems we can design a controller to give, in a sense, the "best possible response". For instance, the ideal response we can expect to a unit step input is 0 at $t = 0$ (when the step is applied) and 1 at all subsequent sampling instants $t = T, 2T, 3T, \ldots$. Hence there is no overshoot and the output reaches the required steady state in a minimal number of sampling instants. Such an output response is known as a deadbeat response. To perform a deadbeat design, the controller is constructed so that it cancels all the unwanted system dynamics, and introduces terms such that a specific closed-loop transfer function is obtained. This is dependent on the type of input to be followed (for example, a step, ramp, etc.). Hence a deadbeat design is tuned for a particular input, and care needs to be taken when a different form of input signal is applied.

To illustrate the design procedure, we use as an example the rigid-body satellite system with $T = 0.05$ s, so that

$$G(z) = 0.00125 \frac{(z+1)}{(z-1)^2} \tag{4.107}$$

If we let our controller $D(z)$ be such that

$$D(z) = \frac{(z-1)}{0.00125\,(z+1)} \tag{4.108}$$

we have exact cancellation leaving just one pole at $z = 1$. The open-loop transfer function becomes

$$D(z)\,G(z) = \frac{1}{z-1} \tag{4.109}$$

which, when used in a unity negative feedback system, as shown in Figure 4.23(a), gives a closed-loop transfer function as

$$\frac{Y(z)}{U(z)} = \frac{1}{z} \tag{4.110}$$

When $U(z)$ is a unit step we have $U(z) = z/(z-1)$ and so

$$Y(z) = \frac{1}{z-1} \tag{4.111}$$

which gives

$$y^*(t) = \delta(t-T) + \delta(t-2T) + \delta(t-3T) + \cdots \tag{4.112}$$

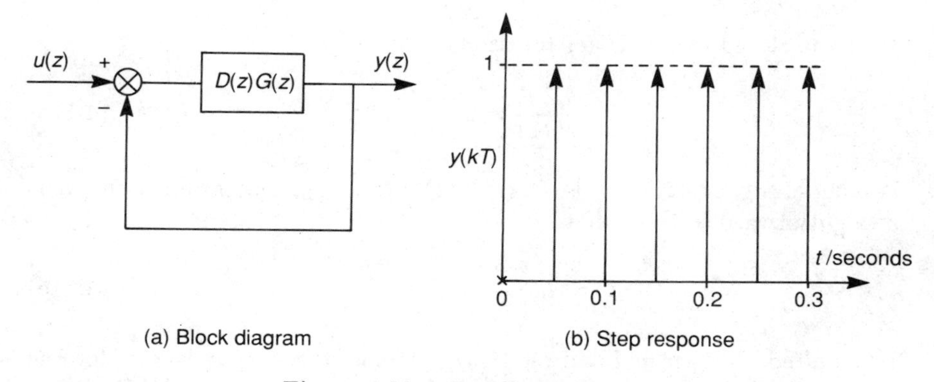

(a) Block diagram (b) Step response

Figure 4.23 Deadbeat design

the deadbeat response as shown in Figure 4.23(b). This satisfies the specifications required.

Hence, for a step input, we need to design the closed-loop transfer function to be $1/z$. It can be shown that for a unit ramp input we should design the closed-loop transfer function to be

$$\frac{Y(z)}{U(z)} = 2z^{-1} - z^{-2} \tag{4.113}$$

This gives the deadbeat response shown in Figure 4.24(a). For a parabolic

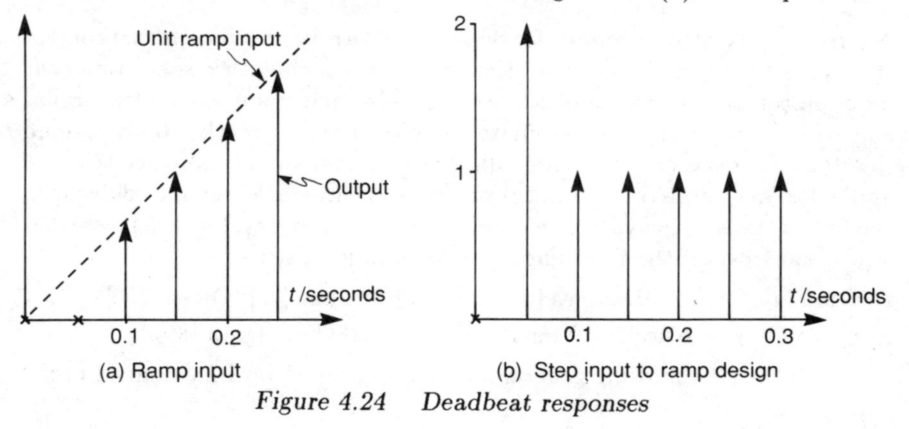

(a) Ramp input (b) Step input to ramp design

Figure 4.24 Deadbeat responses

input the closed-loop transfer function for deadbeat design needs to be

$$Y(z) = 3z^{-1} - 3z^{-2} + z^{-3} \tag{4.114}$$

For further discussion see Kuo [70].

The above-mentioned dependence of the deadbeat design on the input type can be demonstrated by considering, for example, the ramp input case.

Here the closed-loop transfer function

$$\frac{Y(z)}{U(z)} = 2z^{-1} - z^{-2} \tag{4.115}$$

is to be designed. If $U(z)$ is $Tz/(z-1)^2$, that is a unit ramp input, it is straightforward to show that

$$Y(z) = T\left\{\frac{2}{z^2} + \frac{3}{z^3} + \frac{4}{z^4} + \cdots\right\} \tag{4.116}$$

as required, (shown in Figure 4.24(a)), that is, the output is zero for the first two sampling instants and then it follows the input exactly. However, if the input is a unit step, that is, $U(z) = z/(z-1)$, we get

$$Y(z) = \frac{2}{z} + \frac{1}{z^2} + \frac{1}{z^3} + \cdots \tag{4.117}$$

The output sequence is shown in Figure 4.24(b), where we can see an overshoot of 100% at $t = 1$. In some situations this can be quite damaging and so precautions need to be taken when deadbeat designs are implemented in practice.

4.8 Optimal Control Design

Another established technique for designing controllers is via optimal control theory. Here the performance of the design is judged with respect to a cost function (or performance index). Nothing else matters in the optimisation and so it is important to formulate the objectives correctly. Having done this the optimisation procedure will give the control law that needs to be applied to maximise (or minimise) the criteria chosen. We cannot delve into the finer details of the vast area of optimal control theory, but the interested reader can consult the following texts for a fuller discussion:

Continuous Case – Hestenes [42]; Lee and Markus [73]; Owens [90].

Discrete Case – Kuo [70]; Franklin and Powell [35]; Ogata [89].

Consider the following general non-linear discrete optimal control problem:

$$\min_{u} \sum_{k=0}^{(N-1)T} \{\ell(x(k), u(k), k)\} + F(x(NT), NT) \tag{4.118}$$

$$s.t.\quad x(k+1) = f(x(k), u(k), k) \quad \text{for } k = 0, T, \ldots, (N-1)T \tag{4.119}$$

$$x(0) = x_0 \tag{4.120}$$

where x is an n-dimensional state vector, u is an m-dimensional input vector and $s.t.$ stands for "such that". The functions ℓ, F and f are assumed to be continuous in their respective arguments. There may also be other constraints present which restrict further the permissible regions (see Virk [114]). The term $F(x(NT), NT)$ in the objective is the terminal cost, and $f(x, u, k)$ defines the dynamics of the system under consideration. We define the Hamiltonian by

$$H(x(k), u(k), \lambda(k+1), k) = \ell(x(k), u(k), k) + f^T(x(k), u(k), k)\lambda(k+1) \tag{4.121}$$

where the n-dimensional costate vector $\lambda(k)$ is defined by

$$\lambda(k) = \frac{\mathrm{d}H}{\mathrm{d}x}(x(k), u(k), \lambda(k+1), k) \quad \text{for } k = 0, T, 2T, \ldots, (N-1)T \tag{4.122}$$

$$\lambda(NT) = \frac{\mathrm{d}F}{\mathrm{d}x}(x(NT), NT) \tag{4.123}$$

We will assume $T = 1$ for ease of notation. Note that an alternative definition for the Hamiltonian, $H = -\ell + f^T\lambda$, is also commonly used. This definition gives rise to Pontryagin's maximal principle, see Pontryagin *et al.* [95], which is a necessary condition for a control to be optimal. We use the definition $H = \ell + f^T\lambda$ and will subsequently end up with a minimum principle statement. The two approaches are equivalent since

$$\max(f) = \min(-f)$$

and should cause no confusion. A formal statement of the minimum principle now follows:

> If u^* is an optimal control for the above problem, and x^* is the corresponding state trajectory, then there exists a costate function $\lambda^*(k)$ which is the solution of
>
> $$\lambda(k) = \frac{\mathrm{d}H}{\mathrm{d}x}(x^*(k), u^*(k), \lambda^*(k+1), k) \quad k = 0, 1, 2, \ldots, N-1 \tag{4.124}$$
>
> such that
>
> $$H(x^*(k), u^*(k), \lambda^*(k+1), k) = \min_{w \in R^m} H(x(k), w, \lambda^*(k+1), k) \quad \text{for all } k = 0, 1, 2, \cdots, N-1 \tag{4.125}$$
>
> where the minimisation is performed over the feasible controls $w \in R^m$.

Having stated the general non-linear optimisation problem and the minimum principle relating to it, obtaining the numerical optimal solution for a particular application is however quite difficult. The normal procedure is to start with an initial guess for the solution, and then to perform local linearisations about this point so that a simpler optimisation subproblem results, which can be solved more easily. The solution to the subproblem gives a "descent" direction to proceed along to reach the optimal point for the non-linear problem that is being sought. The descent direction together with a step size determines a new, better, estimate of the solution. The problem is then relinearised about this new point and the iteration repeated until a local minimum is found. Since non-linear problems are being dealt with, there may be several local maxima and minima and so several such local optimisations have to be performed before we can ascertain that a global solution has been found. To avoid such practical difficulties we restrict attention to a simpler class of problems where the solution exists in a closed form.

We consider the optimal control of linear dynamical time-invariant systems with respect to a quadratic performance index. This gives rise to the LQP class of optimal control problems which is stated as follows:

$$\min_u \frac{1}{2} \sum_{k=0}^{(N-1)T} \{x^T(k)Qx(k) + u^T(k)Ru(k)\} \quad + \quad \frac{1}{2}x^T(N)Qx(N) \tag{4.126}$$

such that

$$x(k+1) = Ax(k) + Bu(k) \qquad \text{for } k = 0, 1, \ldots, N-1 \tag{4.127}$$

$$x(0) = x_0 \tag{4.128}$$

For a solution to exist, the Q and R matrices need to be positive semi-definite (written ≥ 0), and positive definite (written > 0), respectively. Also, without loss in generality, Q and R can be written so that they are symmetric (see Owens [90]).

Definition: Positive (Semi-) Definite Matrices

(i) An $r \times r$ matrix M is said to be positive semidefinite if

$$x^T M x > 0 \tag{4.129}$$

for all vectors $x \neq 0$.

(ii) Similarly a matrix M is positive definite if

$$x^T M x > 0 \tag{4.130}$$

for all vectors $x \neq 0$.

Q and R are weighting matrices that define the importance of the various states and controls. Note $R > 0$ ensures that the controls do not become too large and so saturation in actuators, etc. is avoided.

The above general minimum principle can be restated for this LQP problem:

> If (u^*, x^*) is an optimal pair for the LQP problem, then there exists a costate function $\lambda(k)$ which is the solution of
>
> $$\lambda(k) = Qx(k) + A^T\lambda(k+1) \qquad (4.131)$$
> $$\lambda(N) = Qx(N) \qquad (4.132)$$
>
> such that

$$H(x^*(k), u^*(k), \lambda^*(k+1), k) = \min_{w \in R^m} H(x^*(k), w, \lambda^*(k+1), k)$$
$$\text{for all } k = 0, 1, 2, \ldots, N-1 \qquad (4.133)$$

The Hamiltonian is defined as

$$H(x(k), u(k), \lambda(k+1), k) = \frac{1}{2}\left\{x^T(k)Qx(k) + u^T(k)Ru(k)\right\} + (Ax(k) + Bu(k))^T\lambda(k+1) \qquad (4.134)$$

To determine the minimising control for the Hamiltonian we must study

$$\frac{\mathrm{d}H}{\mathrm{d}u} = Ru(k) + B^T\lambda(k+1) = 0 \qquad (4.135)$$

Solving this gives the optimal control u^*, and the complete solution to the LQP problem can be obtained by solving the difference equations for the costate backwards from $\lambda(N)$ at time equal to N. This so-called two-point boundary value problem (TPBVP) can be solved explicitly by using the sweep method devised by Bryson and Ho [18], where we assume that the costate is related to the state by a $n \times n$ time-varying gain matrix P, that is, we have

$$\lambda(k) = P(k)x(k) \qquad (4.136)$$

Under this assumption we have, from equation (4.135), that

$$Ru(k) = -B^TP(k+1)x(k+1) \qquad (4.137)$$
$$= -B^TP(k+1)(Ax(k) + Bu(k)) \qquad (4.138)$$

Hence

$$u^*(k) = -(R + B^TP(k+1)B)^{-1}B^TP(k+1)Ax(k) \qquad (4.139)$$

Again, using equation (4.136) in equation (4.131) we get

$$\begin{aligned} P(k)x(k) &= Qx(k) + A^T P(k+1)x(k+1) & (4.140) \\ &= Qx(k) + A^T P(k+1)(Ax(k) + Bu(k)) & (4.141) \end{aligned}$$

Using the optimal $u^*(k)$ we get

$$\begin{aligned} P(k)x(k) &= Qx(k) + A^T P(k+1)Ax(k) \\ &\quad -A^T P(k+1)B\left(R + B^T P(k+1)B\right)^{-1} B^T P(k+1)Ax(k) \end{aligned} \quad (4.142)$$

Hence we have

$$P(k) = Q + A^T[P(k+1) - P(k+1)B\left(R + B^T P(k+1)B\right)^{-1} B^T P(k+1)]A \quad (4.143)$$

as a matrix difference equation (the *matrix Riccati equation*) that can be solved backwards from time N. The boundary condition is given by (4.132), so that $P(N) = Q$.

Having obtained $P(k)$ we can obtain the optimal control

$$\begin{aligned} u^*(k) &= -\Delta(k)x(k) & (4.144) \\ \text{where } \Delta(k) &= (R + B^T P(k+1)B)^{-1} B^T P(k+1)A & (4.145) \end{aligned}$$

Hence if the time period N is known, the above approach can be used to compute $P(k)$, and then control the system through

$$\begin{aligned} x(k+1) &= Ax(k) + Bu^*(k) & (4.146) \\ \text{where } \quad u^*(k) &= -\Delta(k)x(k) & (4.147) \end{aligned}$$

The situation is shown in Figure 4.25. For optimal computer control appli-

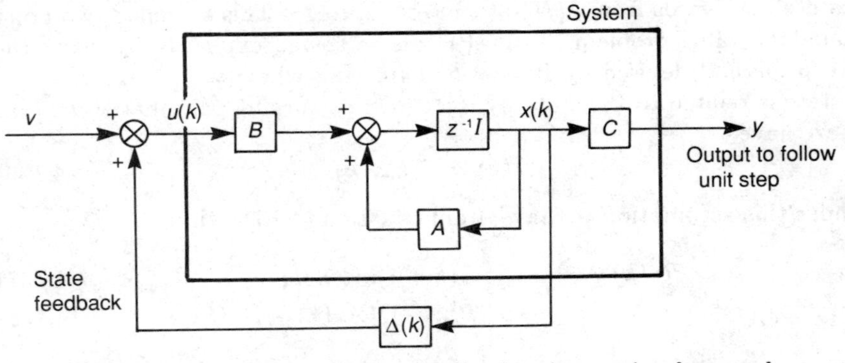

Figure 4.25 Time-varying feedback using optimal control

cations, the optimal control needs to be calculated for all values of time. To determine these values, we can either solve the infinite horizon case when $N \to \infty$ (see Ogata [89]), or consider the finite time interval of N as discussed above. When one interval ends, a new optimal control problem for the next interval can be formulated and solved.

Let us illustrate the design using our rigid-body satellite example. The state-space equations of this system have been shown to equal

$$x(k+1) = \begin{bmatrix} 0 & 1 \\ -1 & 2 \end{bmatrix} x(k) + \begin{bmatrix} 0 \\ 0.00125 \end{bmatrix} u(k) \quad (4.148)$$

$$y(k) = \begin{bmatrix} 1 & 1 \end{bmatrix} x(k) \quad (4.149)$$

where k represents the number of 0.05 s steps. The performance index needs to be formulated to take into account the required objectives. Essentially, these objectives (from before) are that when u is a unit step the output should not overshoot too much and should settle to the correct steady state as quickly as possible (in under 1 s). In the above formulations we are minimising a positive quadratic function and so everything is forced to zero since this is the minimum value. If the required steady state is not zero (as in our case), we must take account of this in the formulation. The situation is similar to that already discussed in section 4.5, where we needed to introduce the concept of servo systems. For our design to proceed along these lines here, it is necessary to extend the above optimal control results to the servo system setting, or modify the problem. We will take the latter option, and we will design an optimal control law to drive the system from a non-zero initial condition to the origin in the state-space. For a discussion on optimal control of servo systems see Ogata [89].

Letting the initial conditions be $x_1(0) = x_2(0) = 0.5$, we have $y(0) = 1$, and the optimal controller will be designed to drive this to zero. The precise solution will depend on the values of the elements of Q (a 2×2 matrix) and R (a scalar), which dictates the weightings on the states and control respectively. If $Q = \begin{bmatrix} 1 & 0 \\ 0 & 1 \end{bmatrix}$, $R = r = 0.001$ and $N = 40$ (an optimising horizon of 2 s). Equation (6.49) can be solved backwards from $P(40) = \begin{bmatrix} 1 & 0 \\ 0 & 1 \end{bmatrix}$ to give

$$P(39) = \begin{bmatrix} 2 & -2 \\ -2 & 6 \end{bmatrix}, \quad P(38) = \begin{bmatrix} 6.94 & -9.9 \\ -9.9 & 18.8 \end{bmatrix}, \text{ etc.}$$

Using these values we can work forwards and calculate the optimal control using equation (4.144) and the corresponding optimal state trajectories using equation (4.146). The output response is shown in Figure 4.26 where it can be seen that the settling requirements are satisfied (for a $\pm 5\%$ tolerance). The performance can be further improved by repeating the design using a larger value for the Q matrix elements and/or a smaller value for r.

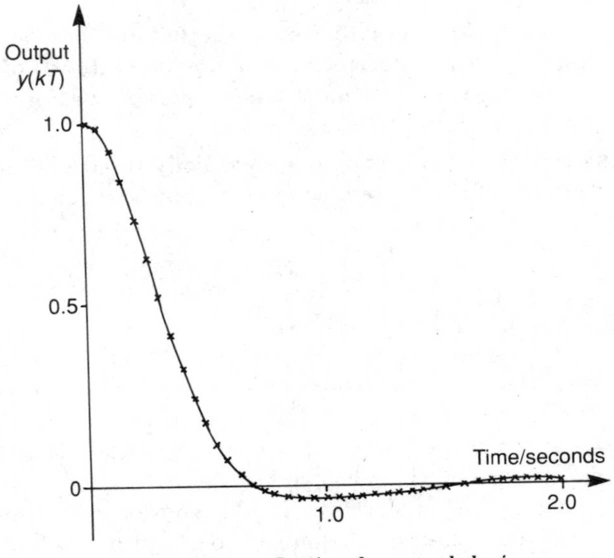

Figure 4.26 Optimal control design

4.9 Controller Design in the Presence of Noise

All the controller designs thus far have considered only the deterministic situation where everything in the system is known to a good degree of accuracy. Hence the system transfer function and/or state-space matrices are defined as are all the signals in the system. Under these conditions all the analysis and designs can be calculated precisely, as discussed in the previous sections. In practice the situation can however be quite different. The signals can be random in nature, as in for example a control valve being set at some level, but in practice it oscillates about this level because of deadband non-linearities. Hence the controlled variables could be varying about a mean level instead of being at a well defined set value. Other disturbance effects (known or unknown) can affect the system performance. The general situation is shown in Figure 4.27 where we can see that noise can enter the system at several places. With these stochastic effects as they are termed, it is no longer possible to define the system behaviour exactly, since all the signals will be corrupted. We can, however, use statistical techniques and the properties of the noisy signals to determine what the responses are *likely* to be. This essentially means we can define mean levels and a range over which the values will vary. Figure 4.28 shows the standard deviation or dispersion for a normally distributed random variable, see Papoulis [91]. Under stochastic conditions our state-space equations become

$$x(k+1) = Ax(k) + Bu(k) + v(k) \qquad (4.150)$$

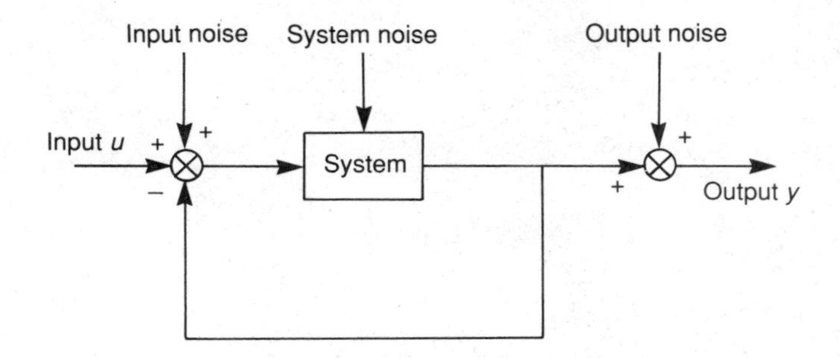

Figure 4.27 Control systems in the presence of noise

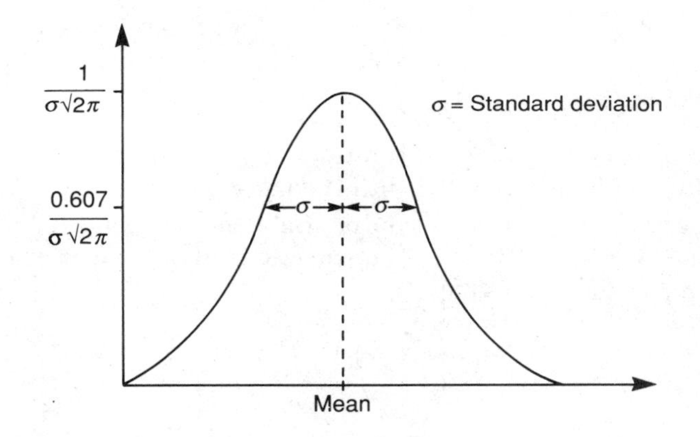

Figure 4.28 Distribution of a normal random process

$$y(k) \quad = \quad Cx(k) + w(k) \tag{4.151}$$

where v and w are random variables representing noise processes affecting the system, with v being the process noise and w being the observation noise. A detailed discussion of the analysis under these conditions is beyond the scope of the text and we can only afford a brief introduction to the area. The interested reader is referred to an excellent discussion by Goodwin and Sin [40] for further details. It is important to realise that for any analysis to be possible, knowledge of probability, random variables and stochastic processes is required; for information on these aspects see, for example, Papoulis [91].

We make the following assumptions:

Assumption 1

$v(k)$ and $w(k)$ for $k = 0, 1, 2, 3, \ldots$, are zero mean stationary white noise processes with covariance given by

$$E\left[\begin{bmatrix} v(i) \\ w(i) \end{bmatrix} \begin{bmatrix} v^T(j) & w^T(j) \end{bmatrix}\right] = \begin{bmatrix} S_v & S_{vw} \\ S_{vw}^T & S_w \end{bmatrix} \delta(i-j) \tag{4.152}$$

where δ is the Kronecker delta function defined by

$$\begin{aligned} \delta(i-j) &= 0 \qquad \text{for } i \neq j \\ &= 1 \qquad \text{for } i = j \end{aligned} \tag{4.153}$$

Assumption 2

The initial state $x(0) = x(k_0)$ is a random variable of mean $\bar{x}_0$ and covariance $S(0)$ and is uncorrelated with the noise processes v and w.

Using these assumptions we can study the behaviour of the system of equations (4.150) and (4.151). Looking at the state equations we have, by taking expected values, that

$$E\left[x(k+1)\right] = E\left[Ax(k) + Bu(k) + v(k)\right] \tag{4.154}$$

or

$$\bar{x}(k+1) = A\bar{x}(k) + B\bar{u} \tag{4.155}$$

$$\bar{x}(0) = \bar{x}_0 \tag{4.156}$$

where the overbar signifies mean values. Now defining the covariance of the states at time k as $S(k)$ we have

$$S(k) = E\left[\left(x(k) - \bar{x}(k)\right)\left(x(k) - \bar{x}(k)\right)^T\right] \tag{4.157}$$

and so we can determine how this varies by analysing

$$\begin{aligned} S(k+1) &= E\left[\left(x(k+1) - \bar{x}(k+1)\right)\left(x(k+1) - \bar{x}(k+1)\right)^T\right] \\ &= E\left[A\left(x(k) - \bar{x}(k)\right)\left(x(k) - \bar{x}(k)\right)^T A^T\right] + E\left[v(k)v(k)^T\right] \end{aligned} \tag{4.158}$$

Since $v(k)$ is uncorrelated with $x(k)$ and $\bar{x}(k)$ we have

$$S(k+1) = AS(k)A^T + S_v \tag{4.159}$$

$$S(k_0) = S(0) \tag{4.160}$$

Similar analysis of the output equation gives its mean as

$$\bar{y}(k) = C\bar{x}(k) \tag{4.161}$$

and its covariance as

$$S_y(k) = CS(k)C^T + S_w \tag{4.162}$$

Hence to define each variable we need two equations – one to define the mean value, and another to define the covariance (or spread) due to the stochastic effects.

Various methods can be used for designing control systems which are subject to such noisy influences, such that the disturbance effects are minimised, thereby allowing the deterministic effects (that is, the mean levels) to dominate. Objectives such as minimising the variance (see Goodwin and Sin [40]) or generalised versions have become classic approaches and are in common use, as are Kalman filters for estimating state vectors under noisy conditions (see Ogata [88]). Optimal control problems and the other design methods discussed in earlier sections can also be formulated for the stochastic case (see Fleming and Rishel [32]).

4.10 Summary

We have used a simple example to demonstrate various controller design methods that can be used in sampled data systems. For most of the approaches discussed, there are no hard and fast rules, and no rigid procedure to follow which will give the exact specifications sought. The situation is stated more in terms of guidelines and compromises that need to be made for satisfying conflicting objectives. The design procedure, in general, needs to be iterated several times, making adjustments each time so that the compromises and achieved performance are optimised. This has been shown in most of the example designs in this chapter.

4.11 Problems

1. Consider the following 2 input/1 output discrete system.

$$\begin{aligned} x(k+1) &= \begin{bmatrix} -0.5 & 3 & 4 \\ 0.5 & -1 & -2 \\ -0.5 & -1 & 0 \end{bmatrix} x(k) + \begin{bmatrix} -2 & 2 \\ 1 & -1 \\ 0 & 1 \end{bmatrix} \begin{bmatrix} u_1(k) \\ u_2(k) \end{bmatrix} \\ y(k) &= \begin{bmatrix} 0 & 1 & 1 \end{bmatrix} x(k) \end{aligned}$$

   (i) Deduce whether the system is stable, state controllable and/or state observable.

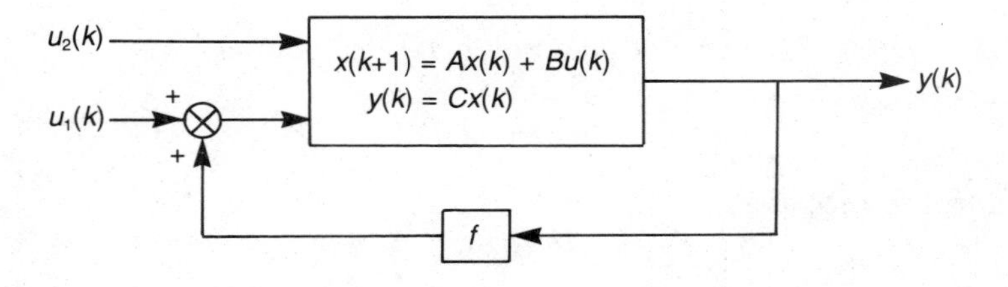

Figure 4.29 Output feedback system

(ii) The output of the system is fed back with respect to the input $u_1(k)$ to form a closed-loop system as shown in Figure 4.29. Determine whether the system can be stabilised in this way and if so calculate a suitable value of f.

2. Consider a unity negative feedback control system with open-loop transfer function $G(s)$, where

$$G(s) = \frac{1}{(s+1)(s+15)}$$

It has been decided to use computer control techniques so that the closed-loop system is as shown in Figure 4.30. Design a controller $D(z)$ so that the performance of the compensated system satisfies the following criteria

settling time, $t_s \leq 1$ second,
damping ratio, $\zeta \geq 0.5$,
zero steady-state error to a step.

You should choose a suitable sampling interval T. Show how your design $D(z)$ would be implemented in practice and evaluate the performance of your design.

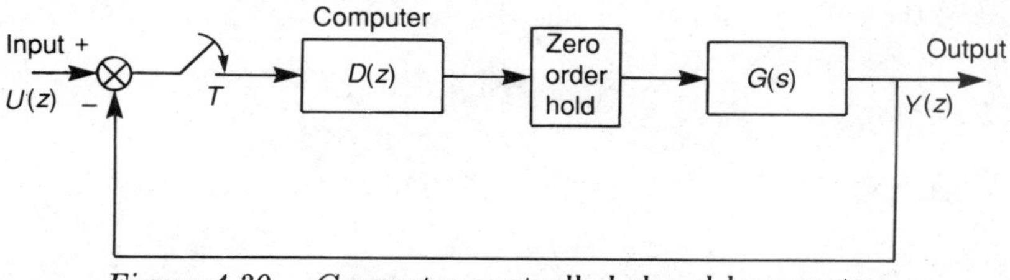

Figure 4.30 Computer controlled closed-loop system

3. Obtain the control sequence to solve the following optimal control problem

$$\min_{u} \sum_{k=1}^{3} \frac{1}{2} \left\{ w_1 x^T(k) Q x(k) + w_2 u^2(k) \right\}$$

such that

$$x(k+1) = \begin{bmatrix} 1 & 2 \\ -3 & 1 \end{bmatrix} x(k) + \begin{bmatrix} 1 \\ 2 \end{bmatrix} u(k)$$

$$x(1) = \begin{bmatrix} 10 \\ 15 \end{bmatrix}$$

when $w_1 = w_2 = 1$ and $Q = \begin{bmatrix} 2 & 0 \\ 0 & 4 \end{bmatrix}$.

If the weight w_2 is changed to zero, explain (without performing detailed calculations) how the optimal solution changes and why.

4. Figure 4.31 shows a block diagram of a computer controlled magnetic suspension system where a coil is energised to provide lifting force. It is required to write a computer program so that the overall closed-loop system satisfies the following specifications

   settling time, $t_s \leq 1$ second,
   damping ratio, $\zeta \geq 0.5$.

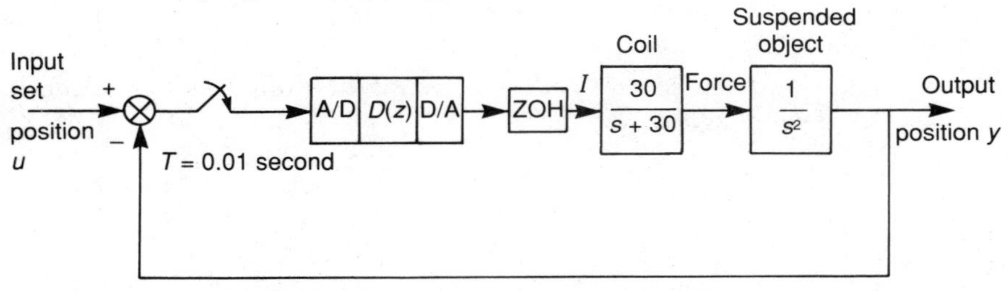

Figure 4.31 Magnetic suspension system

   Without performing detailed calculations, design a suitable continuous controller $D(s)$ and hence obtain the digitised controller $D(z)$. Explain how this $D(z)$ can be implemented in the computer.

5. Consider the linear time-invariant discrete system

$$x(k+1) = \begin{bmatrix} -1 & -2 & -2 \\ 0 & -1 & 1 \\ 1 & 0 & -1 \end{bmatrix} x(k) + \begin{bmatrix} 2 \\ 0 \\ 1 \end{bmatrix} u(k)$$

$$y(k) = \begin{bmatrix} 1 & 1 & 0 \end{bmatrix} x(k)$$

   Find a feedback gain vector so that all the eigenvalues of the resulting closed-loop system are zero. Show that for any initial state, the zero-input response of the compensated system becomes identically zero for $k \geq 3$. This is the *deadbeat* control design.

6. Consider the system shown in Figure 4.32. By determining the continuous state-space representation and digitising it, show that the system

can be described by

$$x\left((k+1)T\right)=\begin{bmatrix}0.5 & 0\\ 0 & 0.25\end{bmatrix}x\left(kT\right)+\begin{bmatrix}-0.5\\ 0.75\end{bmatrix}u\left(kT\right)$$

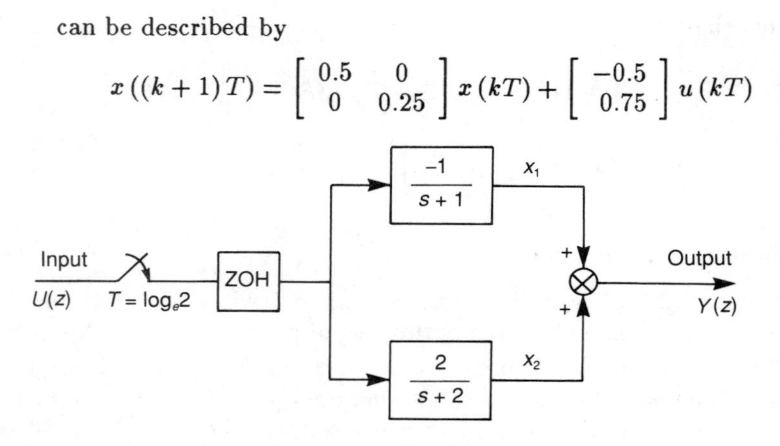

Figure 4.32 Modal control block diagram

Hence determine the output $y(kT)$ for $k=0,1,2,3$ and $k=\infty$ when the input is a unit step. Check your result by using the z-transform method.

7. (i) Assess the controllability, observability and stability of the following digital system:

$$\begin{aligned} x(k+1) &= \begin{bmatrix}-1 & 0.8\\ 0.5 & -1.6\end{bmatrix}x(k)+\begin{bmatrix}1\\ 2\end{bmatrix}u(k)\\ y(k) &= \begin{bmatrix}1 & 0\end{bmatrix}x(k)\end{aligned}$$

(ii) It is required to use optimal control theory to design a feedback strategy over a finite time interval $0\le k\le N$ using a quadratic performance index of the form

$$\min_u \frac{1}{2}\sum_{k=0}^{N-1}\left\{x^T(k)Qx(k)+ru^2(k)\right\}+\frac{1}{2}x^T(N)Qx(N)$$

where Q is a 2×2 positive semi-definite symmetric matrix, $r>0$, and $x(0)=\begin{bmatrix}10\\ 5\end{bmatrix}$. Without performing detailed calculations, show that a suitable controlling feedback signal is obtained by using state-feedback and time-varying optimal gains.
You may assume that a quadratic function

$$J(u)=a_1+a_2u+\frac{1}{2}a_3u^2$$

for $a_3>0$ has a unique minimum at $u^*=-a_2/a_3$, and

$$J(u^*)=a_1-\frac{1}{2}\frac{a_2^2}{a_3}$$

(iii) The state vector needed in part (ii) for the feedback loops is inaccessible. Construct a reduced-order observer for the system and show how the complete state vector can be obtained and used for the application of the optimal control results.

8. Consider the digital control system shown in Figure 4.33. Use the root-locus method to design a discrete compensator $D(z)$ so that the closed-loop system satisfies the following specifications:

   settling time, $t_s \leq 0.4$ second,
   damping ratio, $\zeta \geq 0.7$,
   zero steady-state error to a step.

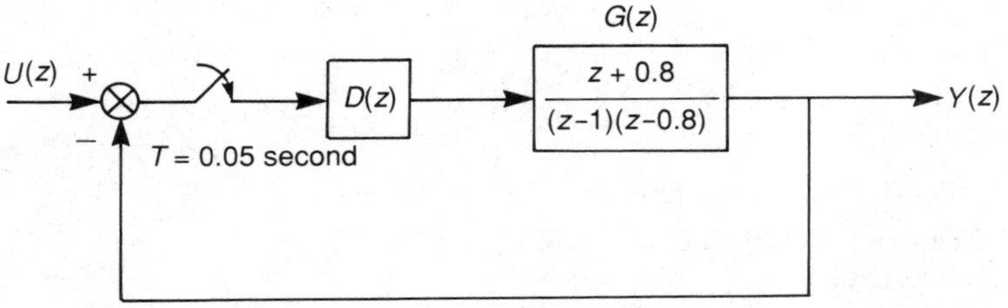

Figure 4.33 Discrete closed-loop system

You should explain the methodology and the reasons for your decisions as the design proceeds.
Assess the performance of your design when a unit step is applied to the system.

9. Figure 4.34 shows a block diagram of a third-order multivariable system with 2 inputs and 2 outputs.

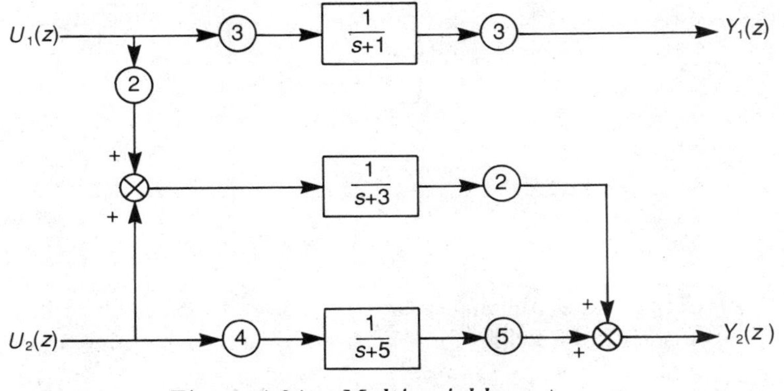

Figure 4.34 Multivariable system

(i) Determine a continuous state-space representation of the system.
(ii) If the inputs U_1 and U_2 are applied using a digital computer and zero-order holds, with a sampling interval of 0.1 s, calculate the discrete state-space form for the system.

(iii) Calculate the time response for the first 0.3 s of the digital system when U_1 and U_2 are unit step functions and initial conditions are zero.

10. Show that a SISO system with transfer function

$$\frac{Y(z)}{U(z)} = \frac{z+0.3}{z^2-0.6z-0.16}$$

can be represented in state-space form by

$$\begin{aligned} x(k+1) &= \begin{bmatrix} 0 & 1 & 0 \\ 0 & 0 & 1 \\ 0.08 & 0.46 & 0.1 \end{bmatrix} x(k) + \begin{bmatrix} 0 \\ 0 \\ 1 \end{bmatrix} u(k) \\ y(k) &= \begin{bmatrix} 0.15 & 0.86 & 1 \end{bmatrix} x(k) \end{aligned}$$

Explain fully your answer.

11. Consider the following second-order SISO system

$$\begin{aligned} x(k+1) &= \begin{bmatrix} -1 & -0.5 \\ -1 & -1.5 \end{bmatrix} x(k) + \begin{bmatrix} 1 \\ 2 \end{bmatrix} u(k) \\ y(k) &= \begin{bmatrix} 0.5 & 0 \end{bmatrix} x(k) \end{aligned}$$

(i) It is required to feedback the state vector as shown in Figure 4.35 so that the resulting closed-loop system has zero steady-state error to a unit step input. Determine whether this is possible and if so calculate a suitable feedback vector F.

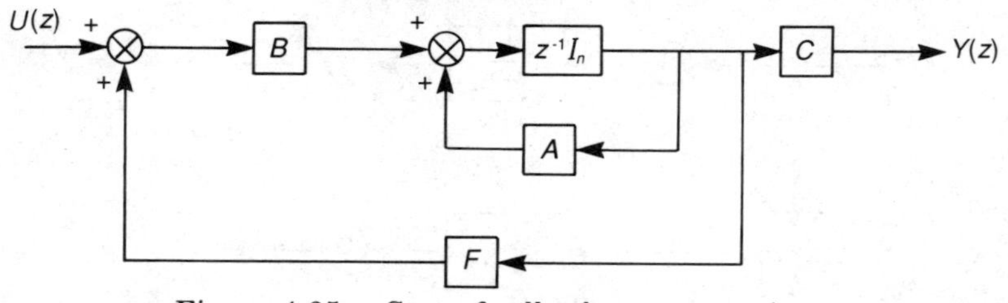

Figure 4.35 State-feedback compensation

(ii) If the states are inaccessible at the output, can the objectives in (i) still be achieved? Discuss how and determine an observer that satisfies the required objectives.

12. Figure 4.36 shows a block diagram of a second-order mechanical system in a state-space representation.

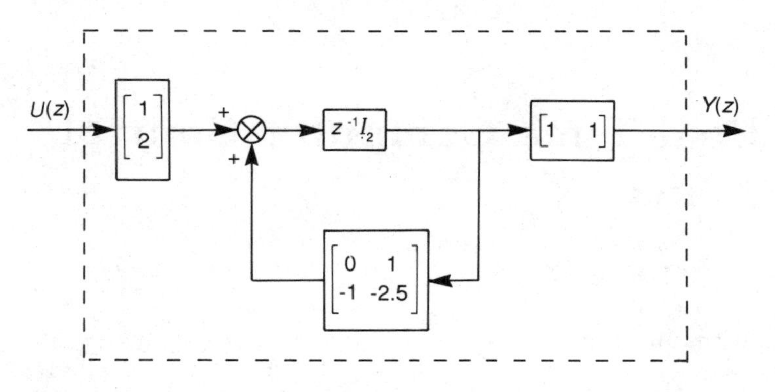

Figure 4.36 Discrete state-space block diagram

It is required that the system be compensated using state feedback so that the closed-loop system has poles at $z = 0.4 \pm j0.3$. However the complete state is not directly available at the output for feedback purposes and state estimation is necessary. Show how this can be done and determine suitable feedback and observer matrices. You should estimate only those state components that need to be estimated.

5 Real-Time Computer Control Systems

5.1 Introduction

Up to now we have concentrated mainly on z-domain mathematical analysis methods to provide the reader with a sound understanding of sampled data systems and how such systems can be designed. After the design phase, the implementation of the solutions poses difficulties which as yet have not been considered in this text. In this chapter we tackle these issues since the use of computers in controlling real systems on-line has become standard practice, and consequently control engineers need to be aware of the important factors. The material is more descriptive than earlier chapters, but it is felt that such a discussion is necessary so that a clear insight into real-time computer control systems may be gained.

The main point to remember is that the computer, in these cases, works in conjunction with some process from which data is gathered and processed with subsequent application of control signals to the plant. This occurs in an endlessly repeated cycle. Also, all parts of the process work at rates dictated by their dynamics, and so possess their own inherent time scale requirements – these may be of the order of milliseconds, hours, days or even years. It is important to realise that the computer needs to be linked directly into these time scales of the process for effective operation. Therefore we must select and program the computer in such a way that it can keep track of the passage of "real-time" independently of its own internal operations. Our own familiar system of seconds, minutes, hours, etc. normally becomes the standard of reference by which the computer initiates actions relating to the process under control.

The concept of "real-time" also implies the ability of the computer to respond to stimuli from the system in a timely fashion, that is, it must respond sufficiently fast in order to accommodate the needs of the process. For example, if an emergency condition arises in the plant and is signalled to the computer, the computer must be capable of reacting to the requirements of the process sufficiently fast in order to handle the emergency. Clearly, for simple processes operating on long time scales (hours or days), the design of a real-time computer control system will not involve any critical timing problems, and practically any digital computer could be selected for the job. For complex processes (involving information), or for processes operating on short time scales (milliseconds or seconds), the challenge of real-time responses become much more critical, requiring careful attention in the selection of the computer and in designing the overall system.

Although we may attempt to generalise real-time computer control concepts, it must be remembered that each such system is in fact a unique creation. Hence the computer user, despite his main concern with his process and what he requires to do with it, is forced to pay some attention to the structure of the computer system, to the computer/process interface, and to programming fundamentals if he wishes to design and to make best use of the computer/process system. In the past, the design of real-time systems could only be performed well if the user was familiar with many details of the computer hardware, and could perhaps even program in the machine language itself, in order to be able to program the computer efficiently. Things have improved significantly since these early days and we are in fact still in the midst of an explosion of technology.

Several trends are emerging – the rapid development of inexpensive, fast computer hardware, the introduction of standard high-level languages (with necessary extensions for real-time applications), and the wider availability of sophisticated executive systems for monitoring computer operations – all of which are reducing the requirement for specialised knowledge. Computer hardware and software systems now on the market require the real-time user to understand only a few additional commands whose use is relatively straightforward. This sort of progress was expected and necessary for the widespread application of computer control techniques. Even so, it is advisable that a person attempting real-time applications should have a true understanding of the basics of computing in general, and real-time computing in particular, if he is to use the computer in an effective way. Furthermore, despite major efforts to simplify and generalise the interconnection of the process and the real-time computer, the interface between them will always present particular problems which require detailed and specific information for their proper solution.

This chapter represents an attempt to bring together all of the fundamentals of real-time computing necessary for control applications. It is impossible to cover all the areas in great detail but, wherever possible, references will be cited so that the interested reader can follow up on our discussions. Our approach will be divided into the hardware and software design aspects and how the two are brought together, with correct communications and synchronizations, to give the overall solution sought. We start by looking at the hardware set up.

5.2 Hardware Requirements

The Digital Computer and Peripheral Equipment

Although almost any digital computer can be used in a real-time control application, not all adapt to this purpose equally easily. A block diagram of a general purpose digital computer is shown in Figure 5.1 where we can see that it is composed of four main units.

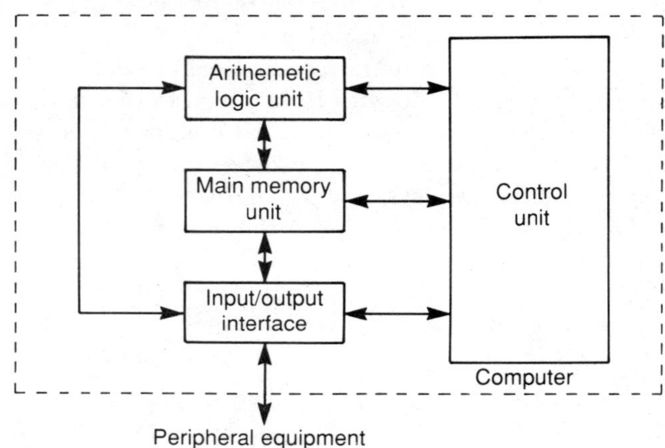

Figure 5.1 Block diagram of a digital computer

The Arithmetic Logic Unit (ALU) contains all the hardware necessary to carry out arithmetic and logic commands, for example, to add two numbers, subtract two numbers, check to see if one number is larger than another number, etc. The ALU contains the working registers of the computer, in which numbers are temporarily stored before they are used in computations. Modern computers can have many registers, which allow greater ease in programming, since numbers can be stored and used in the processing without accessing the main memory, which is usually slower.

All the units in the computer are constantly under the supervision of the Control Unit. This part of the computer is responsible for reading a program from memory, interpreting it, causing the appropriate action to take place (for example, add two numbers in the ALU), selecting the next consecutive program statement for reading, etc. The Control Unit is capable of carrying out "branching" within a program, that is, a change in the normal sequence of operation, when appropriate. This capability is responsible for much of the versatility of the general purpose computer since it allows for interrupts, internally or externally generated, to be accommodated.

The Memory Unit is used for the storage of data and of the computer program itself. Normally the Control Unit carries a sequence of program statements to be executed, stored in consecutive memory locations. The

Main Memory is often referred to as the "fast memory", to distinguish it from "bulk memory" or "slow memory" which is usually a separate, peripheral device.

The final unit in our simplified view of the computer is the Input/Output (I/O) interface which is necessary for the computer to communicate with the outside world, that is, all of its peripheral equipment. Interfaces in most computers consist of a set of bidirectional data lines and control lines, usually referred to as buses, and the logic necessary to detect and respond to external "events". These events usually take the form of a request for some kind of action on the part of the computer, which would then have to interrupt its normal processing. The ability to respond to an external "interrupt" is a requirement for any computer, but it is in fact the very basis of the real-time control application. It is this capability that allows the real-time computer to keep track of the passage of time independently of its normal operations, and to watch multiple processes, each with a different set of demands which must be serviced by the computer. Obviously, the I/O interface is also necessary for reading a program into the computer and for transferring results out of the computer. As with the other computer units, the operations of the I/O interface are normally co-ordinated by the Control Unit.

The term "Central Processing Unit" (CPU) is ordinarily used to refer to a combination of the Arithmetic and Control Units. In any computer control system design, the choice of the hardware reduces to the selection of the CPU, since this performs the actual processing, and therefore defines the computing capacity of the system. Features of the CPU that are important in this selection include

(i) the word length; this defines the integer range representable and hence defines the precision of the calculations. In addition, the word length also defines the amount of memory directly accessible with one instruction word (2^n for an n-bit word), although it can be increased by multiple word operations, but with a resulting time penalty;

(ii) the instruction set, addressing methods and number of registers which defines how easy, flexible and fast the CPU is to program;

(iii) the information transfer rates, defining the transfer rates within the CPU, and between the CPU and external interfaced devices;

(iv) the interrupt structure, of vital importance in any real-time control application, and needing to be of a flexible and multi-level nature.

Having looked at the digital computer itself we can go on to a discussion on the peripheral equipment in a typical real-time computer control set up. The equipment used can be divided broadly into two categories, the first of which is related to the computer and user-related tasks, and the second of which handles the needs of the process under study. A typical situation is shown in Figure 5.2.

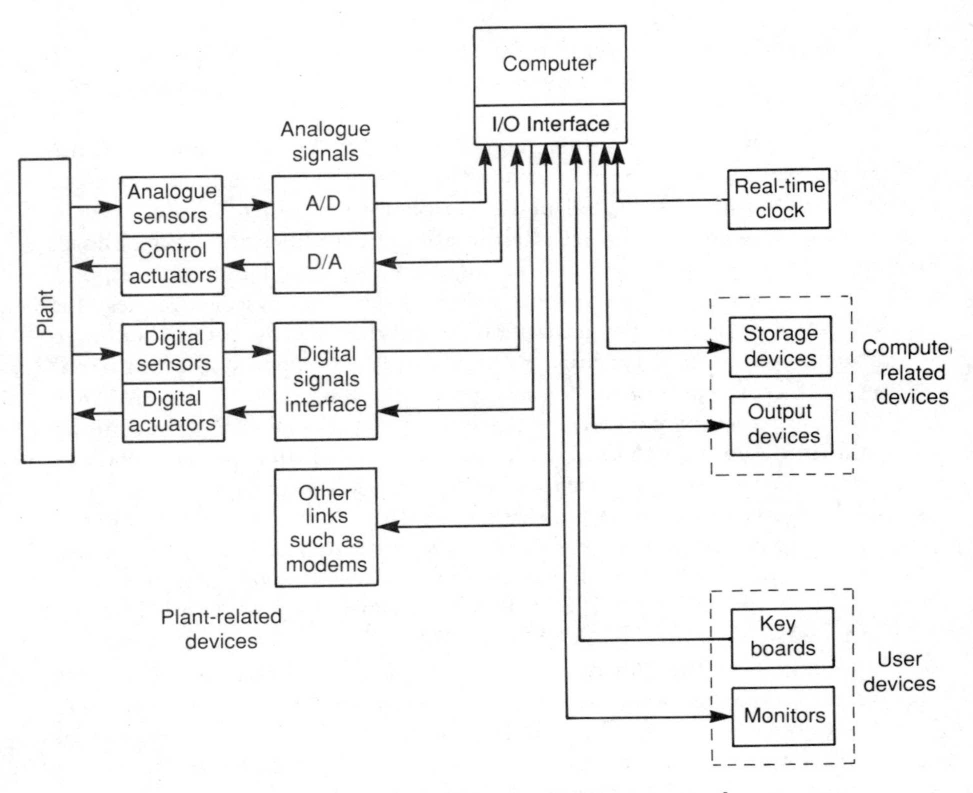

Figure 5.2 Typical real-time computer control system

The Real-Time Clock

An important peripheral device shown in Figure 5.2 is the real-time clock, which is not necessarily a clock at all, in the normal sense of the word. In many systems it is merely a pulse generator (usually crystal driven) with an accurately controlled frequency. The pulses are fed to the computer via the I/O interface so that the computer, when correctly programmed, can use them to ascertain the passage of time. For example, if the clock interval is one second, a counter inside the computer can be set to zero at the start of an experiment and incremented by one count each time the clock pulse interrupts the processor. This gives a continuous estimate of time in seconds from the start of the experiment. By resetting a counter to zero exactly at midnight, one could easily program a routine to give the time on a 24 hour basis.

The choice of the basic clock interval (that is, the clock precision) has to be a compromise between the timing accuracy required and the load on the CPU. If too small an interval, that is, high precision, is chosen then the CPU will spend a large proportion of its time simply servicing the clock and will not be able to perform much other useful processing work.

The real-time clock based on the use of an interval time and interrupt-driven software suffers from the disadvantage that the clock stops when the power is lost and needs to be restarted with the current value of the real-time. Real-time clocks are now becoming available in which the clock and date function are carried out as part of the interface unit, hence the unit acts like a digital watch. Real-time can be read from the device and the unit can be programmed to generate an interrupt at a specified frequency. These units are usually supplied with a battery back-up so that even in the absence of mains power the clock function is retained.

The real-time clock interrupts are also used to multiplex the CPU so that the computer can perform several tasks seemingly at the same time. In fact, the CPU, being a sequential machine, can only perform one task at a time but by dividing its attentions to different operations for short time intervals (defined by the clock interrupts) several tasks can be computed in rotation (but together). This technique is known as time-slicing.

Interface Units

As shown in Figure 5.2 there can be many types of units connected to the computer. The analogue/digital interface units (already discussed in chapter 1) convert analogue process signals, such as temperatures, speeds, stresses, etc., via transducers into digital form for use in the computer (A/D), and digital actuating values into analogue control signals for application to the process (D/A). Other forms of interface devices for plant/computer junctions are necessary for

(i) *digital quantities:* these can be either binary, such as a switch which can be on or off, or a generalised digital quantity, such as the output from a digital instrument in binary coded decimal form;

(ii) *pulses or pulse trains:* some instruments, for example, flowmeters, provide output in the form of pulse trains and clearly need to be handled;

(iii) *telemetry:* the increasing use of remote outstations, such as electricity substations and building energy management outstations have increased the use of telemetry. The data may be transmitted by landline, radio or the public telephone network.

Further discussions on these process-related interfaces are given in Bennett [13] and Mellichamp [83]. Other peripheral devices needed for computer or user related tasks include

(a) *interactive devices or terminals:* such devices enable direct communication between the user and the computer system, and hence it is possible manually to input data, programs and commands at run time as well as to interrupt the computer from performing its normal sequence of operations;

(b) *computer related input/output devices:* these include devices such as line printers, card readers and punches, data loggers and other peripherals used for repeated input/output of large quantities of data;

(c) *auxiliary storage equipment:* these devices extend the storage capabilities of the computer and include devices such as magnetic disks, drums and tapes.

Further discussion on these peripherals can be found in Mellichamp [83].

5.3 Software Aspects

Having discussed the hardware devices needed to support real-time computer control applications, we turn our attention to the software aspects so that information can be communicated and the operations performed in an efficient and effective manner. A program for translating and implementing the designed controller obviously needs to be programmed so that the process variables are gathered into the computer and the controller strategies applied. In addition to this, many other tasks need to be performed as part of the overall design. These include

(i) ensuring correct data transfer between the devices;

(ii) making sure faults and emergencies are handled adequately and that the process performance remains within specified limits;

(iii) updating operator displays and responding to manual inputs;

(iv) logging data reports for inspection.

In this section we will outline how these, plus other related aspects, are dealt with and the types of options available to the designer.

Controller Implementation

Assume that a controller $D(z)$ has been designed, by using any of the methods discussed in chapter 4, so that the required closed-loop system is as shown in Figure 5.3. The next step is the actual implementation of the design so that the z-domain transfer function $D(z)$ is realised by a computer algorithm. The conversion is relatively straightforward as we now show. Let us suppose that the controller designed has the form

$$D(z) = \frac{M(z)}{E(z)} \tag{5.1}$$

where M is the input signal applied to the system as shown, and E is the

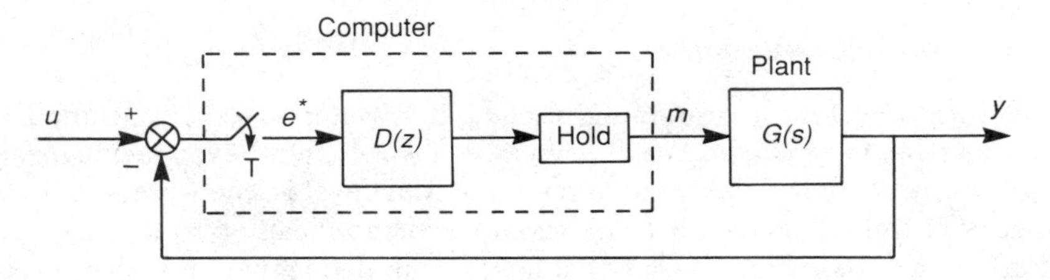

Figure 5.3 Computer controller implementation

error. Assuming these are both polynomials in z, we have

$$\frac{M(z)}{E(z)} = \frac{b_0 z^n + b_1 z^{n-1} + \cdots + b_{n-1} z + b_n}{z^n + a_1 z^{n-1} + \cdots + a_{n-1} z + a_n} \tag{5.2}$$

Dividing the numerator and denominator by the highest power in z, that is z^n in our case, we get

$$\frac{M(z)}{E(z)} = \frac{b_0 + b_1 z^{-1} + \cdots + b_{n-1} z^{-n+1} + b_n z^{-n}}{1 + a_1 z^{-1} + \cdots + a_{n-1} z^{-n+1} + a_n z^{-n}} \tag{5.3}$$

Now since multiplication by z^{-1} is equivalent to a delay of T seconds we can convert the expression into the time domain by writing $z^{-1}e^*(t) = e^*(t-T)$, etc. Hence we obtain

$$\begin{aligned} m^*(t) &= b_0 e^*(t) + b_1 e^*(t-T) + \cdots + b_n e^*(t-nT) \\ &\quad -a_1 m^*(t-T) - a_2 m^*(t-2T) - \cdots - a_n m^*(t-nT) \end{aligned} \tag{5.4}$$

Thus the input that needs to be applied can be generated simply by using stored previous values of the errors and inputs.

Although we have illustrated the technique for a proper $D(z)$ (where the order of the numerator and denominator are the same), it is impossible in practice to implement the design exactly since it requires the measurement of $e^*(t)$, and the calculation and application of $m^*(t)$, to take place instantaneously. The technique may work adequately for slow systems where processing delays are small but it is normal practice to insert a delay of at least one sampling interval so that $m^*(t)$ depends on errors up to the last sampling instant at time $(t-T)$. A simple way of ensuring this is by designing controllers that are strictly proper (where the numerator order is at least one less than the denominator order). Continuing with this line of reasoning it is obviously impossible to implement a non-proper $D(z)$ (where the order of the numerator is larger than the order of the denominator), since $m^*(t)$ will depend on future errors. Therefore in the design procedure we only consider realisable controllers.

Programming Languages

The lowest level of programming in computer systems is the machine language itself which is defined by binary coded words that have specific meaning for a particular CPU. Programming at this level is "non-user-friendly" since coded numbers of this form have little or no meaning to the majority of users and consequently lead to difficulties in debugging. For this reason it is rarely done. The next level of programming is in the so-called assembly language, where the binary coded word instructions are mapped on a direct statement-for-statement basis into another mnemonic coded form that is easier to understand. For example "LDA, NUMB" could be used to represent an instruction to transfer a number stored in memory to a register instead of the machine language equivalent "0011000100100100". A few examples of the Intel 8080 assembly language instructions are (see Kuo [70]) as follows:

Mnemonic	*Description*	*Machine code*
ADD M	Add M to A	10000110
JMP	Jump unconditional	11000011
IN	Input	11011011
XRA M	Exclusive-OR M with A	10101110

where A is a register, and M represents the main memory. Clearly, writing programs in an assembly language is much easier than writing them in the machine language. The computer itself converts the assembly instructions into machine instructions by means of a program called the assembler. On a higher programming level still are the high level languages which are even easier to use and require little or no knowledge of the lower level workings of the computer. Hence, since most computer applications users are non-specialists in computer science, most of the programming is carried out in one of these high level languages. The main exceptions to this are when the user is forced to program in assembly because of severe time specifications, as well as limited memory resources, or when a non-standard interface device needs to be used in a control loop.

There is a whole variety of high level languages available for use, including Fortran, Pascal, C, Modula 2, Ada, Basic and Forth. The language selected in practice depends on a number of factors such as the computer to be used, features required and the previous experience of the designer. The standard Basic and Forth languages are interpreter based whereas the others are compiler based. The distinction is that in the latter we have to run the written program through another program called a compiler that translates the high level instruction such as "Print ⋯ " into the (possibly) hundreds of low level machine commands that are needed by the CPU to perform the required task (printing). The successfully compiled program can then be executed (run). Interpreter based programs on the other hand are not compiled but are loaded together with an interpreter program and run. As each command is reached it is interpreted and executed. Both have advantages and disadvantages which can be summarised as follows:

(a) Since interpreter based programs do not need compiling, it is easy to modify programs by simply changing a line or so and re-running. The price paid is that since each instruction is interpreted at run time, it takes appreciably longer to run a standard Basic program than it does for, say, a Fortran one.

(b) Compiler based programs pose difficulties at the development phase since each time the code is amended it needs to be recompiled before the edited program can be run. This can be tedious, but once developed, the compiled program gives better performance because it is in executable form.

Other application orientated languages also exist, such as APT (for numerically controlled machine tools) and CUTLASS (a language for process control), but these are rather specialised developments and outside the scope of this book (see Bennett [13]).

Another important program that the real-time user ought to be aware of is the "Operating System" which is responsible for supervision and co-ordination of all programs that are run on the CPU. It is therefore responsible for supporting and controlling all basic activities, high level language support, I/O access, etc. (see for example Hoare and Perrott [44]).

5.4 System Design Co-ordination

In the preceding sections we have discussed various hardware and software aspects that real-time control engineers need to take into consideration when implementing a controller design. However there are other important issues he must also bear in mind if the final design is to perform its tasks in a satisfactory way. The main areas of concern include efficient transfer of information between the computer and the devices, adequate interrupt design, and ensuring that the operations are sequenced properly. In addition, effects of the discretisation procedure need to be kept under close scrutiny so that the digital system performs adequately. Two considerations in this latter regard are the sampling rate selection, and a check on the quantisation errors, which we discuss next.

Sampling Rate Selection and Quantisation Errors

The sampling theorem (stated in chapter 1) gives a lower bound on the sampling rate as twice the highest signal frequency of interest in any particular application. However (as already discussed) for smooth and accurate control performance, much higher rates than this are usually employed. Clearly the performance of digital control systems improves as the sampling rate is increased. The precise figure depends on the particular application and the sort of performance levels required, but the following factors need to be taken into consideration in the decision making process.

(i) Financial considerations; sampling faster generally means increasing hardware costs since faster and more powerful computers are necessary to compute the controller calculations in the smaller sampling intervals. In addition A/D, D/A converters need to be able to operate at the sample rate selected, and the cost of these also grows with increasing speeds.

(ii) Allowing for the time delays introduced in the control loop as a result of sampling, and conversion times for A/D and D/A. These delays can accumulate and hence add significant destabilising phase lags.

(iii) The smoothness of the controlled response required.

(iv) The regulation effectiveness as measured by the response errors due to stochastic and other disturbance effects.

(v) The sensitivity to system parameter variations; such variations cause differences in the model and actual system parameters to occur, and give rise to errors in the system response characteristics. These errors grow with increasing sampling period.

(vi) Measurement errors and the influence of analogue pre-filters or anti-aliasing filters.

These points are discussed in more detail in Houpis and Lamont [46], and Franklin *et al.* [35]. The first point suggests using the slowest sampling rate that is permissible, whereas the other factors suggest using the fastest rate possible. Obviously a compromise needs to be reached, and in practice, for most computer control systems applications, the sample rates used are $10 - 20$ times the highest signal frequency of interest (see Katz [64] or Franklin *et al.* [35]).

The finite word lengths used in digital computers to represent system parameters and data values mean that quantisation errors are introduced because all the numbers are rounded to a LSB. These digital numbers are operated upon mathematically in the computer algorithms, and so the errors propagate through the system. To ensure that such quantisation errors do not become excessive, it is necessary to develop error models (see Houpis and Lamont [46]; Franklin *et al.* [35]), so that error analysis of the rounding-off can be performed. With these models it is possible to study various aspects of quantisation effects, including

(a) worst-case error which gives an upper bound on the error resulting from the rounding-off of the data values;

(b) steady-state worst case, where transient errors are ignored as long as they die out, and an upper bound on the steady-state error is obtained;

(c) effects of parameter round-off.

Limit cycle phenomena also arise in digital systems owing to the quantisation of signals. Such oscillatory motions are due to the non-linear aspects of the digital controller based upon finite word length and the associated processing. In these cases a dead band exists within which the system output cannot be controlled because of the resolution of the LSB (see Houpis and Lamont [46]; Franklin *et al.* [35]).

The word lengths used in the digital computer, A/D and D/A define the resolution used in representing the numbers, and hence dictate the control performance achievable. The longer these are, the tighter the error tolerances, but with increasing cost penalty. Computers with 16 bit and 32 bit word lengths are commonly used nowadays, with I/O interfacing units of 8 – 16 bits depending on performance specifications demanded.

Computer/Device Communications

As we have seen in Figure 5.2 several forms of devices need to be connected to the computer, and for effective control performance, information needs to be transferred between them and the computer. Each device has its own controller (interface) that permits the communication to take place. The simplest technique is known as unconditioned transfer where the computer carries out the operation whenever the particular section of code is executed. Although this can work for fast devices, slower ones may not be able to supply or receive information at the rate requested and data can be lost. Buffers can be included to overcome some of these difficulties but buffers can become full and data can still be lost. A better method is obtained if the status of the device is checked before the transfer is made. This method can work well but the major disadvantage of this conditional transfer approach is that the computer may need to wait (and waste) an appreciable amount of time before the device is ready. Of course, the computer could be programmed to carry on with another task if the device is not ready and retry the transfer at a later time. In fact the programmer must ensure that the device is regularly accessed so that all the data transference is carried out.

These problems are eliminated if the computer has the capability of responding to an interrupt from any of the devices connected to it. With such a facility, it is possible for the computer to divert from its current program and respond to the interrupt. Having serviced the request of the interrupt, the computer can return to its previous operation and continue from where it left off.

Interrupt Handling

A simple method of incorporating an interrupt facility is to introduce an interrupt line, which is regularly monitored by the CPU. If any device wishes to interrupt the computer, it can activate this interrupt line. The CPU, sensing the interrupt, branches to a separate section of the operating system known as the Interrupt Handler (IH), which temporarily stores all the information necessary for later resumption of operations from where the CPU was interrupted. The IH then determines which device caused the interrupt and branches to the appropriate subroutine for servicing that device. After this has been done the IH restores the CPU to its state at the time of interrupt, reinitialises the interrupt line and returns to execution of the interrupted program.

For the above simple interrupt handler to function we need to ensure that a second device cannot interrupt the IH itself, and must wait until the initial interrupt routine is completed. That is, we need to disable subsequent interrupts as soon as the IH becomes activated. Although such a philosophy may function adequately, the approach is rather inefficient since some devices require fast response while others are less critical. A better solution is obtained if a priority structure is introduced, with process related activities at the highest priority (level 0); the next level (1) could be high-speed computer-related I/O data storage; (2) could be the real-time clock, and so on. Hence, having selected a suitable priority structure, the hardware can be set up so that an interrupt from a device on one level locks out all devices on lower levels until it is serviced completely; but interrupts from a device on a higher level priority would be permitted. Many types of complicated interrupt handling can be designed to operate under any conceivable situation. For a detailed discussion, see Mellichamp [83], but the essential feature is that the IH itself needs to be able to be interrupted by devices on a higher priority (that is, those not masked). In these cases the handler needs to store the status of the interrupted program. The status data cannot overwrite the previous status data but is appended onto a "stack" that grows with the number of interrupts. Having serviced the highest priority interrupt the IH can work its way down the stack completing all the unfinished interrupts.

Multi-tasking and Background/Foreground

Stacking interrupts as just discussed allows us to interleave the operations of many or all external devices, so that the computer can perform several tasks simultaneously by responding to interrupts which indicate that a device is ready for the next piece of information. Multi-tasking, as it is known, is the process by which the program is structured into several tasks (each given a priority) that can be run in parallel. A task scheduler then takes the responsibility of scheduling when each task is performed, depending upon its priority and its readiness to be executed.

An extension of this priority multi-tasking, commonly referred to as background/foreground, is often used in real-time control applications where the computer needs to interact continuously with the process. All the tasks that need to be performed can be divided into real-time and non-real-time jobs. The real-time tasks are put on the high priority (foreground) and the non-real-time on the low priority (background). In this way, when foreground operations are described, the processor is switched over to the usual task scheduler for implementation of all the foreground jobs. When there is no foreground activity required, the background task scheduler is allowed to run and perform all the non-critical jobs, such as batch computing, and update operator displays. Whenever a foreground activity needs to be performed, the foreground task scheduler takes over again.

5.5 Summary

In this chapter we have given a broad overview of computer systems in control applications and how discrete controller designs can be implemented. In addition, the aspects a design engineer needs to take into consideration when selecting the hardware and software necessary for implementing his design, have been discussed.

6 Advances in Computer Control

6.1 Introduction

Since the early computer control applications of the 1960s, computational concepts and techniques have always played a major role in the area of control engineering. This role has in fact grown in importance over the years because of the increasing complexities of the control tasks considered. These tasks required the corresponding increases which have occurred in the sophistication of the computing hardware, design methods and software tools available. The advances in computer technology over the last twenty years or so have been so dramatic that it is difficult to appreciate the magnitude of the achievements. It has been said that if the car industry had experienced the same level of growth, cars would be able to achieve fuel economies of 250,000 miles per gallon of petrol while cruising at 500,000 miles per hour! Such revolutionary improvements in computer systems have allowed sweeping changes to be made in a wide spectrum of applications. On the whole the changes have been to give better performance and more flexible solutions. In this book we have been mainly interested in the effect of these developments on control engineering, which also have been significant. In particular, as discussed in chapter 1, the introduction of the microprocessor and its use as a low cost computing element has had a profound effect on the way in which the design and implementation of the control system is carried out, and to some extent, on the theory which underpins the basic design strategies.

Two significant areas of current change have been due to the initiatives for the Japanese "Fifth Generation Computers" and include consideration of

(i) artificial intelligence, and
(ii) parallel computing algorithms and architectures.

Much research work is taking place not only in these areas but also in the fields of data/information, communication networks and human–computer interaction. These areas could also have a major effect on computer control applications and so we have included discussions on such futuristic aspects, even though their use at the moment is limited. We start by giving a brief introduction to artificial intelligence in computer systems.

6.2 Artificial Intelligence

In contrast to the precise numeric or algorithmic methods described thus far, artificial intelligence (AI) is a branch of computer science that uses symbolic processing and heuristic techniques, to produce computer programs that operate in a manner that could be described as "intelligent". That is, programming computer systems in a way which allows decisions to be reasoned from encoded rules and knowledge, without specific information being available. In simpler terms, AI has often been described as the study of how to make computers do things at which, at the moment, people are better at doing. All the various activities that together form the discipline of AI have several factors in common. They all involve problems that are usually complicated and difficult to define mathematically, yet are amenable to computer representation and are solved by applying various combinations of human expertise, experience and intuition.

The area of AI is littered with jargon and, although the ideas are relatively straightforward, it is difficult for the newcomer easily to understand and get to grips with what is being said. We will try to minimise the jargon in what follows but some of the terminology is difficult to avoid.

We now turn to an important breakthrough to emerge from the field of AI, namely expert systems, or knowledge-based systems.

Expert Systems

Expert systems are computer programs which are able to represent and reason about some application area, and to solve problems and give advice about that domain. The more formal definition of an expert system is that it is the embodiment within a computer of a knowledge-based component from an expert skill, in a way which offers intelligent advice or takes intelligent decisions about that knowledge domain.

Since the mid 1960s the field of AI has achieved considerable success in the development of expert systems, and today there are many such systems, for example:

(i) *MYCIN* (see Shortliffe [103] and Davis *et al.* [24]) for diagnosing and treating diseases;

(ii) *PROSPECTOR* (see Duda *et al.* [30]) used in prospecting for minerals;

(iii) *R1* (see McDermott [78], [79]) for configuring computer systems;

which can give performance levels equal to, or even better than those of the human expert. Such performances are possible because an expert system is generally constructed using knowledge from many experts and so can prove to be better than an individual expert. An expert system usually consists of three main components, as shown in Figure 6.1.

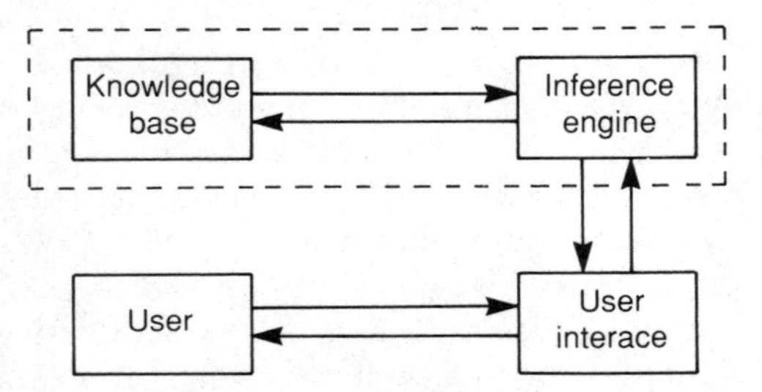

Figure 6.1 Block diagram of an expert system

The part of the expert system that contains the representation of the knowledge, called the knowledge base, is usually quite separate from the part of the system which reasons about the knowledge, called the inference engine. Such a physical separation of the two main parts of an expert system allows the knowledge base to be easily changed or added to. In addition, it also allows the inference engine to be used at different times with different knowledge bases.

The area of expert systems has concentrated on the knowledge that underlies human expertise, rather than on formal reasoning methods, thereby reducing the importance of domain independent problem solving theory. There are several reasons for doing this, which include the following:

(a) There are many problems that do not have precise algorithmic solutions; for example, in control applications it is not possible to state *a priori* which design method is the most appropriate, and therefore it cannot be defined in a way which is sufficiently precise for the needs of current methods of symbolic and mathematical reasoning. Current mathematical techniques would find it very difficult to perform tasks such as representing knowledge, describing abstract problems, and the bringing of diverse sources of knowledge to bear on a problem.

(b) Humans become experts in a particular field because they are knowledgeable. Therefore for computer systems to reach the same levels of expertise, they too should use the same knowledge.

(c) The process of representing expert knowledge on a computer can substantially reduce the costs of knowledge reproduction and exploitation. This in turn speeds up the process of knowledge refinement by allowing private knowledge to become open to public test and evaluation.

For knowledge to be represented in a computer, it must first be extracted from the expert. Knowledge elicitation has often been described as the narrowest bottleneck (see Barr and Feigenbaum [12]) in the process of expert system construction, because there is, as yet, no established methodology for achieving it completely satisfactorily. There are many methods currently being employed, ranging from simple question—answer interviews, to complex psychological techniques such as protocol analysis, multidimensional scaling and concept sorting, see for example Hayes-Roth *et al.* [41]. A precise description of these techniques does not concern us here, suffice to say that they allow expert knowledge to be elicited without being implicitly stated by the expert. This is a very important criterion for knowledge elicitation, because simply asking an expert to state what he/she knows can never adequately capture all the knowledge that is actually known or employed in the knowledgeable decision-making process. In fact it may be true to say that experts quite often are not aware of how much they know or how they use their knowledge in reaching a decision.

Once knowledge has been elicited, it has then to be encoded onto the computer in such a way that allows the user to understand what the system is doing, and why. Suitable matching of the encoded knowledge and user expectations is greatly facilitated by employing the concept of mental models, see for example Young [119] and Norman [87]. The term *mental model* is used to describe a cognitive mechanism for representing and drawing conclusions about a system, which is built by the user while interacting and learning about the system. The mental model helps the user to understand how the system works, and to predict its behaviour. Thus, by investigating the mental model of the particular knowledge area that typical users possess, the knowledge can be encoded to ensure that the expert system employs the same model to represent the knowledge. If this is achieved then the system will appear to the user to be employing the same or similar control strategies that seem to be the most natural to the user.

We illustrate these and other important features of expert systems by considering a few examples.

Examples of Expert Systems

One of the first expert systems was MYCIN, which was developed at Stanford University to help doctors diagnose and treat infectious diseases. A typical consultation is now given:

MYCIN: Have you obtained positive cultures indicating general type?

DOCTOR: Yes

MYCIN: What type of infection is it?

DOCTOR: Primary bacteremia

MYCIN: Let's call the most recent positive culture CULTURE-1. From what site was CULTURE-1 taken?

DOCTOR: From the blood

MYCIN: Let's call the first significant organism from this blood culture ORGANISM-1. Do you know the identity of ORGANISM-1?

DOCTOR: No

MYCIN: Is ORGANISM-1 a rod or a coccus or something else?

DOCTOR: Rod

MYCIN: What is the gram stain of ORGANISM-1?

DOCTOR: Gramnegative

MYCIN: Did ORGANISM-1 grow aerobically?

DOCTOR: Yes

MYCIN: Did ORGANISM-1 grow anaerobically?

DOCTOR: Yes

MYCIN: What is the suspected portal of entry of ORGANISM-1?

DOCTOR: The gastrointestinal tract

⋮ ⋮

When MYCIN has sufficient information for it to be able to identify the disease and recommend treatment, it begins to ask questions that are relevant to the treatment that it is about to suggest:

MYCIN: My recommendation will be based on the opinion that the identity of ORGANISM-1 may be 1 Pseudomonas-aeruginosa, 2 Klebsiella-pneumoniae, 3 E. Coli, 4 Bacteroides-fragilis, 5 Enterobacter, or 6 Proteus-non-mirabilis. On a scale of 0 to 4, where higher numbers indicate greater severity, how would you rate the patient's degree of sickness?

DOCTOR: 3

MYCIN: Does the patient have a clinically significant allergic reaction to any antimicrobial agent?

DOCTOR: No

MYCIN: How much does the patient weigh?

DOCTOR: 70 kilograms

MYCIN: My first recommendation is as follows: In order to cover for items 1, 2, 3, 5, and 6, give Gentamycin using a dose of 119 mg (1.7 mg/kg) q8h IV [or IM] for 10 days. Modify dose in renal failure. And in order to cover for item 4, give Clindamycin using a dose of 595 mg (8.5 mg/kg) q8h IV [or IM] for 14 days. If diarrhoea or other gastrointestinal symptoms develop, check for pseudo-membranous colitis. Do you wish to see the second choice?

DOCTOR: No.

Another well known expert sytem called PROSPECTOR (Duda *et al.* [30]) was developed to help field geologists. It advises on the presence of certain deposits and it selects the best place for drilling sites. It is quite famous in expert system circles for having discovered a previously unknown deposit of molybdenum.

Both MYCIN and PROSPECTOR are concerned with areas where the knowledge is often inexact, and so both systems have had to develop a method to reason with this uncertainty. MYCIN uses a set of measures of belief called "certainty factors" which are influenced by the theory of confirmation. PROSPECTOR uses conditional probabilities and Bayes' Theorem.

A third example, which is well documented, is R1 which is used in computing a working configuration of components from a list of system requirements for the VAX-11 computer system. This is not an easy task, as the systems usually consist of 50–150 components. The main ones are the central processor, memory control units, unibus and massbus interfaces, all connected to a synchronous backplane. Many different peripheral devices, such as printers and disk drives, can be supported by the bus, and so many configurations are possible. R1 decides if a configuration is consistent with the information that it holds about individual components.

The rules of MYCIN and PROSPECTOR produce a measure of certainty, and so these two systems may be classified as probabilistic expert systems. R1 however has IF–THEN rules that lead to absolute determination. R1 can therefore be classified as a deterministic expert system.

6.3 Knowledge-Based Systems in Control Engineering

Having given a brief introduction to AI and expert systems it is apparent that such knowledge-based systems could play a major role in areas where human expertise is scarce, that is, there are not enough human experts to satisfy requirements, or where no established theory exists, and practitioners rely on knowledge and intuition. Our own area of control engineering is one where AI could have a dramatic impact. With the appropriate expert system, a design engineer could be capable of developing designs of the sort discussed in earlier chapters of this book without initially being aware of the detailed information that they require to develop such designs. Steps towards the development of AI systems for control applications (see Leitch [75]) have already begun. The main questions being tackled are where the expertise is put and what expertise is used. Two distinct approaches are emerging in this respect, namely

(a) *Expert Control* – the expertise is put on top of the existing control system designs and takes corrective action if anything goes wrong. The expertise is essentially the knowledge and skill of the control engineer on-line, see Astrom *et al.* [6].

(b) *Knowledge-Based Control* – knowledge about the process itself is used to derive the controller actions. The knowledge is usually expressed as a set of rules, and consequently the approach is also known as rule-based control and includes fuzzy logic controllers discussed below. The knowledge in this case is the skill and intuition of the process operator, and the expertise is placed in the feedback path.

Conventional Controller Design Methods

The analytical design methods presented in earlier chapters are based on the assumption that good mathematical models for physical systems can be developed which accurately represent their physical behaviour. The models are then used in the determination of the precise controller strategies to be implemented on the physical system. Such methods can give good results as already demonstrated but it is the intention here to outline their main shortcomings and show how AI techniques can offer alternative solutions. The main problem with the conventional control system design methods is that the existence of the model is too strong an assumption – in many processes an adequate representation is difficult, if not impossible, to obtain over the complete operating range. In practice even if a process can be modelled, the time and effort required to extract the model, coupled with the shortage and expense of the necessary expertise, combine to render the analytical approach non-viable. If the modelling is not performed correctly, poor performance and loss of confidence in the system results. Limitations such as these have been improved by recent research into adaptive or self-tuning control algorithms which estimate on-line parameters of a model of assumed structure for the process under study. The model is then used in the design of the controller needed to achieve the required specifications. Although improving the situation, the performance of these self-tuning systems is still heavily dependent upon the assumptions of system order, time delays, system structure, etc. Another major failing with these analytical methods is that the human understanding of the operation of the process and its mathematical description is often alien. Therefore a man–machine interface problem exists since meaningful explanations of control decisions cannot be given to operators, who cannot implement their knowledge of the process to fine-tune the controller.

Manual Control Methods

In contrast to these sophisticated mathematical techniques we also have manual control methods which can be shown to give good results. If a human is used as a controller in the feedback loop, his decisions are based upon general and vague hypotheses about the (possibly complex) process, and used to infer suitable control actions. The inference is usually symbolic in nature and rarely numeric. For instance an operator may know that increasing the input causes an increase in the output, and vice versa, and that only this rule need be used in determining the control actions.

A disadvantage of manual methods is that they are only really useful for slow processes. Since the performance depends on a human, it can vary significantly owing to various factors such as stress, boredom and tiredness. So, although the methods work, they can be improved significantly if they could be automated. This is the goal of Knowledge-Based Control Systems – to combine the "vague" symbolic processing capabilities of humans with the speed, capacity and alertness of machines.

6.4 Fuzzy Sets and Rule-Based Control

Fuzzy sets were introduced by Zadeh [120] to incorporate concepts such as vagueness into computer systems. In conventional set theory an element either belongs to a set or does not, but in normal human decision making this sharp distinction does not normally exist. For example, to define someone as being "intelligent" is meaningless since there is no definite IQ above which a person is described as intelligent, that is, there are "fuzzy" levels where he could be referred to as "reasonably intelligent", etc. Such blurring of the set boundaries are obtained in fuzzy set theory by defining a membership function of a fuzzy set which can be between 0 and 1. In traditional set theory the membership function can only take the values 0 or 1.

We present next an introduction to fuzzy sets so that the basic ideas and how they are used in control systems can be appreciated. Our discussion will follow the tutorial paper by Sutton and Towill [106].

In fuzzy set theory, if X is a space of all possible points (universe of discourse), then a fuzzy set A in X is characterised by a membership function $\mu_A(x)$ which associates with each point $x \in X$ a real number in the interval [0,1] that represents the grade of membership of $x \in A$, that is

$$A = ((x, \mu_A(x)) \mid x \in X) \tag{6.1}$$

Thus, the nearer the value of $\mu_A(x)$ to unity, the higher the grade of membership of x. In this way imprecise and qualitative information can be expressed in an exact way. As an example, suppose it is necessary to specify linguistic measures of "*speed in the range 30 miles per hour (mph) to 100 mph*". In addition suppose that such a measure is categorised by "*speeds about 60 mph*". An ordinary set which defines this can be expressed by defining a range (say ±10 mph) and assigning the membership function to be 1 or 0 depending upon whether the speed is within this range or not, as shown in Figure 6.2(a). A fuzzy set of the same situation could have a membership function as shown in Figure 6.2(b).

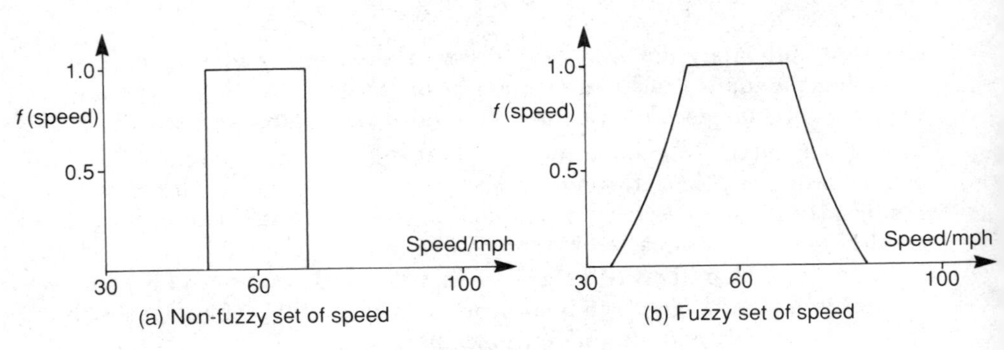

Figure 6.2 Concept of fuzzy sets

The conventional set has a definite boundary which defines membership or non-membership, whereas the fuzzy set allows the qualitativeness of the measure to be reflected in a gradual membership transition. The transition (or membership function shape) is defined by the user and can be chosen to be a variety of shapes. As yet no firm guidelines are available for selecting an optimal form.

In addition to its graphical representation a fuzzy set may be represented mathematically. When X is a finite fuzzy set it may be expressed as

$$A = \frac{\mu_A(x_1)}{x_1} + \frac{\mu_A(x_2)}{x_2} + \cdots + \frac{\mu_A(x_n)}{x_n} \tag{6.2}$$

It should be noted that the "+" sign in equation (6.2) denotes the set theory union operation "∪" rather than the arithmetic sum. Also the notation $\frac{\mu(x)}{x}$ does not denote division, instead it simply relates a particular membership function to a value on the universe of discourse.

If X is continuous then the fuzzy set may be written as

$$A = \int_X \frac{\mu_A(x)}{x} \tag{6.3}$$

Consider a fuzzy set A defined as "*approximately equal to 5*" on a universe of discourse $X = \{0, 1, 2, \cdots, 10\}$. We may describe it as

$$A = \frac{0}{0} + \frac{0.2}{1} + \frac{0.4}{2} + \frac{0.6}{3} + \frac{0.8}{4} + \frac{1}{5} + \frac{0.8}{6} + \frac{0.6}{7} + \frac{0.4}{8} + \frac{0.2}{9} + \frac{0}{10} \tag{6.4}$$

Basic Fuzzy Operations

Fuzzy sets may be operated upon in a similar manner as ordinary sets by means of simple definitions, see Zadeh [120]. Suppose three fuzzy sets of

$X = [0, 10]$ can be labelled *large, medium*, and *small*, and are described by

$$large = \frac{0}{0} + \frac{0}{2} + \frac{0.1}{4} + \frac{0.3}{6} + \frac{0.6}{8} + \frac{1}{10} \tag{6.5}$$

$$medium = \frac{0.3}{0} + \frac{0.6}{2} + \frac{1}{4} + \frac{1}{6} + \frac{0.6}{8} + \frac{0.3}{10} \tag{6.6}$$

$$small = \frac{1}{0} + \frac{0.6}{2} + \frac{0.3}{4} + \frac{0.1}{6} + \frac{0}{8} + \frac{0}{10} \tag{6.7}$$

These sets are shown in Figure 6.3(a).

Union

The union of two fuzzy sets A and B is a fuzzy set C, written $C = A \bigcup B$ or $A + B$, whose membership function is defined by

$$\mu_C(x) = \max\{\mu_A(x), \mu_B(x)\} \tag{6.8}$$

This corresponds to the connective OR. Therefore, from equations (6.5), (6.6) and (6.7):

$$\begin{aligned} large \text{ OR } medium &= \frac{\max[0, 0.3]}{0} + \frac{\max[0, 0.6]}{2} + \frac{\max[0.1, 1]}{4} \\ &\quad + \frac{\max[0.3, 1]}{6} + \frac{\max[0.6, 0.6]}{8} + \frac{\max[1, 0.3]}{10} \\ &= \frac{0.3}{0} + \frac{0.6}{2} + \frac{1}{4} + \frac{1}{6} + \frac{0.6}{8} + \frac{1}{10} \end{aligned} \tag{6.9}$$

Similarly,

$$large \text{ OR } small = \frac{1}{0} + \frac{0.6}{2} + \frac{0.3}{4} + \frac{0.3}{6} + \frac{0.6}{8} + \frac{1}{10} \tag{6.10}$$

The resulting fuzzy sets are shown in Figure 6.3(b).

Intersection

The intersection of two fuzzy sets A and B is a fuzzy set C, written $C = A \bigcap M$, whose membership function is defined by

$$\mu_C(x) = \min\{\mu_A(x), \mu_B(x)\} \tag{6.11}$$

This corresponds to the connective AND operation, and using equations (6.5) and (6.6) gives

$$\begin{aligned} large \text{ AND } medium &= \frac{\min[0, 0.3]}{0} + \frac{\min[0, 0.6]}{2} + \frac{\min[0.1, 1]}{4} \\ &\quad + \frac{\min[0.3, 1]}{6} + \frac{\min[0.6, 0.6]}{8} + \frac{\min[1, 0.3]}{10} \\ &= \frac{0}{0} + \frac{0}{2} + \frac{0.1}{4} + \frac{0.3}{6} + \frac{0.6}{8} + \frac{0.3}{10} \end{aligned} \tag{6.12}$$

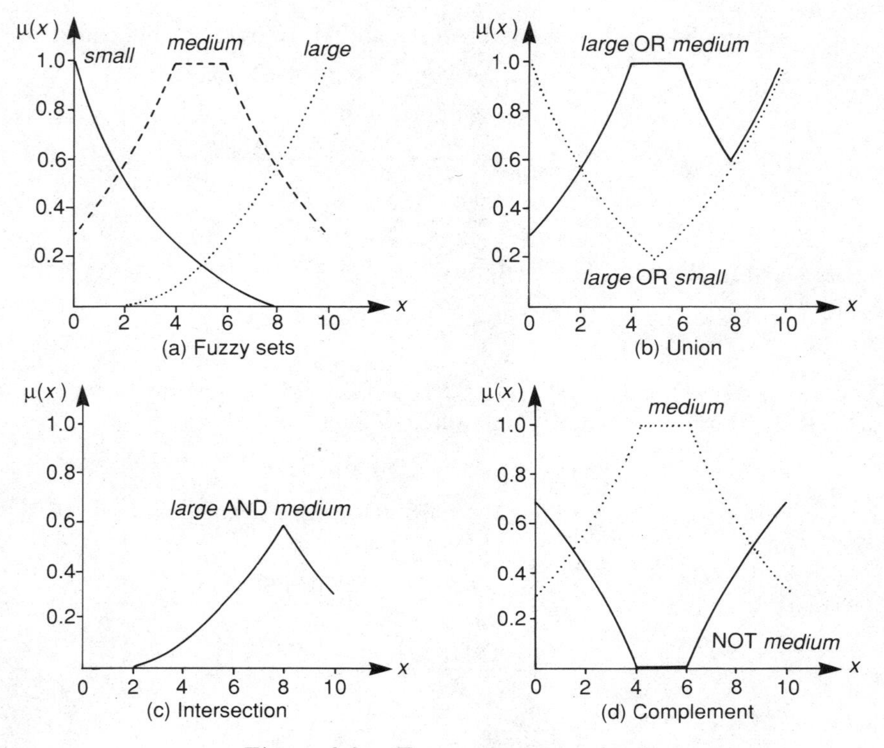

Figure 6.3 Fuzzy sets operations

which is shown in Figure 6.3(c).

Complement

The complement of a fuzzy set A is denoted by $\bar{A}$, with a membership function defined by

$$\mu_{\bar{A}}(x) = 1 - \mu_A(x) \tag{6.13}$$

This corresponds to the negation NOT. Hence, since *medium* is defined by equation (6.6) then

$$\begin{aligned} \text{NOT } medium &= \frac{[1-0.3]}{0} + \frac{[1-0.6]}{2} + \frac{[1-1]}{4} \\ &\quad + \frac{[1-1]}{6} + \frac{[1-0.6]}{8} + \frac{[1-0.3]}{10} \\ &= \frac{0.7}{0} + \frac{0.4}{2} + \frac{0}{4} + \frac{0}{6} + \frac{0.4}{8} + \frac{0.7}{10} \end{aligned} \tag{6.14}$$

as shown in Figure 6.3(d).

It is possible to enhance the description of the fuzzy sets by using qualitative expressions such as "very" and "rather", see Zadeh [121]. As an example, consider a fuzzy set A, then the fuzzy set "*very* A" can be defined as

$$very\ A = \sum_i \frac{\mu_A^2(x_i)}{x_i} \tag{6.15}$$

and "*rather* A" by simply moving along the universe of discourse an amount c, that is

$$\mu_{rather\ A}(x) = \mu_A(x+c) \tag{6.16}$$

Fuzzy Relations

For effective control performance it is necessary to derive the relationship between the system inputs and the system outputs. Such a relationship can be obtained by using a fuzzy conditional statement. A conditional statement between a fuzzy input variable A, and a fuzzy output variable B, on different universes of discourse U and Y respectively, is given by a linguistic implication statement

$$A \rightarrow B \quad \text{or} \quad \text{IF } A \text{ THEN } B \tag{6.17}$$

which reads "{ *input condition* A} implies { *output condition* B }", where A is known as the antecedent and B as the consequent. An example of a linguistic implication relationship for an automobile system, between input fuel, say, and output speed, might be

IF {*input fuel is high*} THEN {*speed is much greater than* 20 *mph*} (6.18)

This implied relation, R, is expressed in terms of the Cartesian product of the sets A and B and is denoted by

$$R = A \times B \tag{6.19}$$

For finite sets, R has its membership function defined by

$$\mu_R(u,y) = \mu_{A\times B}(u,y) = \min[\mu_A(u), \mu_B(y)], \qquad u \in U, \quad y \in Y \tag{6.20}$$

Alternatively, for continuous sets:

$$R = A \times B = \int_{U\times Y} \frac{\mu_A(u) \bigcap \mu_B(y)}{(u,y)} \tag{6.21}$$

Generally, equation (6.21) is in matrix form, therefore

$$R = A \times B = \sum_{U\times Y} \frac{\mu_A(u) \bigcap \mu_B(y)}{(u,y)} \tag{6.22}$$

or

$$A \times B = \begin{bmatrix} \min\left[\mu_A(u_1), \mu_B(y_1)\right] & \cdots & \min\left[\mu_A(u_1), \mu_B(y_n)\right] \\ \vdots & \ddots & \vdots \\ \min\left[\mu_A(u_m), \mu_B(y_1)\right] & \cdots & \min\left[\mu_A(u_m), \mu_B(y_n)\right] \end{bmatrix} \quad (6.23)$$

where $U = \{u_1, u_2, \cdots, u_m\}$ and $Y = \{y_1, y_2, \cdots, y_n\}$.

To illustrate the meaning of these implied relations, consider the following fuzzy conditional statement

$$\text{IF } \{A \text{ is } small\} \text{ THEN } \{B \text{ is } large\} \quad (6.24)$$

where the different universes of discourse U and Y with their corresponding fuzzy sets A and B are given by equations (6.7) and (6.5) respectively. Thus using equation (6.23):

$$\begin{aligned} A \times B &= \begin{bmatrix} \min[1,0] & \min[1,0] & \cdots & \min[1,0.6] & \min[1,1] \\ \min[0.6,0] & \min[0.6,0] & \cdots & \min[0.6,0.6] & \min[0.6,1] \\ \min[0.3,0] & \min[0.3,0] & \cdots & \min[0.3,0.6] & \min[0.3,1] \\ \min[0.1,0] & \min[0.1,0] & \cdots & \min[0.1,0.6] & \min[0.1,1] \\ \min[0,0] & \min[0,0] & \cdots & \min[0,0.6] & \min[0,1] \\ \min[0,0] & \min[0,0] & \cdots & \min[0,0.6] & \min[0,1] \end{bmatrix} \\ &= \begin{bmatrix} 0 & 0 & 0.1 & 0.3 & 0.6 & 1 \\ 0 & 0 & 0.1 & 0.3 & 0.6 & 0.6 \\ 0 & 0 & 0.1 & 0.3 & 0.3 & 0.3 \\ 0 & 0 & 0.1 & 0.1 & 0.1 & 0.1 \\ 0 & 0 & 0 & 0 & 0 & 0 \\ 0 & 0 & 0 & 0 & 0 & 0 \end{bmatrix} \end{aligned} \quad (6.25)$$

It can be seen that the relation matrix shown in equation (6.25) was derived using the Cartesian product of $A \times B$. This relationship can be represented graphically by the surface illustrated in Figure 6.4, where the rows of the matrix equation (6.25) correspond to sections of the surface as shown.

A fuzzy conditional statement can consist of fuzzy sets from more than two disparate universes of discourse. Consider the disparate universes of discourse U, Y and W, with their corresponding fuzzy sets A, B and C respectively, forming:

$$\text{IF } \{A\} \text{ THEN } \{B\} \text{ THEN } \{C\} \quad (6.26)$$

Thus the fuzzy relation is given by

$$R = A \times B \times C = \int_{U \times Y \times W} \frac{\mu_A(u) \bigcap \mu_B(y) \bigcap \mu_C(w)}{(u, y, w)} \quad (6.27)$$

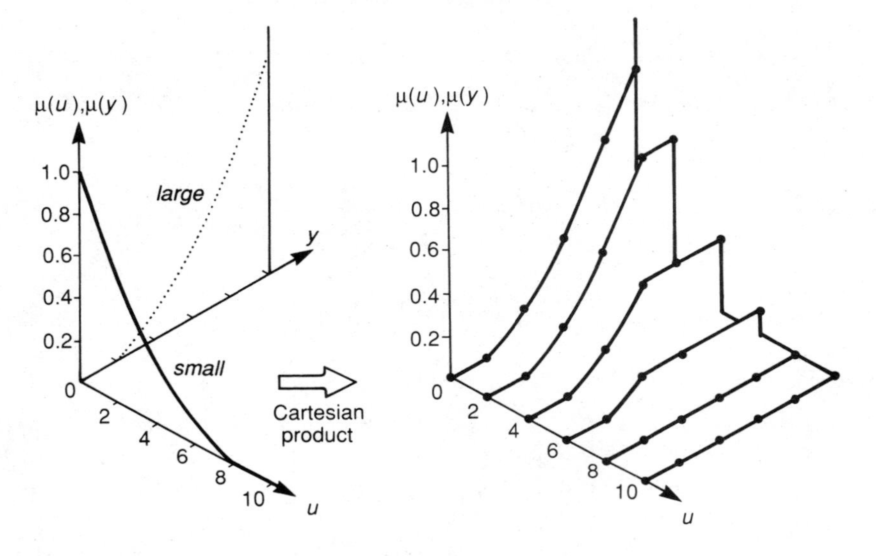

Figure 6.4 *Concept of fuzzy relations*

which can be expressed as a three-dimensional relation matrix by

$$A \times B \times C = \sum_{U \times Y \times W} \frac{\mu_A(u) \bigcap \mu_B(y) \bigcap \mu_C(w)}{(u, y, w)} \tag{6.28}$$

When formulating a control strategy, it is often necessary to form a combination of several fuzzy conditional statements, each of which will have an associated fuzzy set R^i. If this is the case, the individual R^i values are combined to give an overall R by calculating the union of them all. Hence

$$R = R^1 + R^2 + \cdots + R^N \tag{6.29}$$

where R^i is the fuzzy set produced by the ith rule and N is the number of rules. In this way a number of conditional statements can be formed to provide adequate control. It is impractical to have a rule for every conceivable situation and therefore the *composition rule of inference* is used. This allows situations not explicitly covered to be handled by enabling decisions to be inferred from the programmed rules. That is, given a fuzzy relation $R = A \times B$, and a fuzzy set value A' of A, the relation is used to infer the corresponding value B' by using the *composition rule of inference*. The inference of B' is written as

$$B' = A' \circ R = A' \circ (A \times B) \tag{6.30}$$

where "o" denotes the max–min product. Thus the membership function of the output fuzzy subset B' is defined by

$$\mu_{B'}(y) = \max_u \, \min\left[\mu_{A'}(u), \mu_R(u, y)\right] \tag{6.31}$$

To illustrate the inference rule, consider the following logical statement:

$$\text{IF } \{\textit{yaw error greater than } 10^\circ\} \text{ THEN } \{\textit{rudder}, \ \delta, \ \textit{reduced}\} \tag{6.32}$$

where

$$\{\textit{yaw error greater than } 10^\circ\} = \frac{0}{10.0} + \frac{0.2}{12.5} + \frac{0.5}{15.0} + \frac{0.8}{17.5} + \frac{1}{20.0} \tag{6.33}$$

and

$$\{\textit{rudder}, \ \delta, \ \textit{reduced}\} = \frac{1}{-10.0} + \frac{0.8}{-7.5} + \frac{0.4}{-5.0} + \frac{0.1}{-2.5} + \frac{0}{0} \tag{6.34}$$

Using equation (6.23) the relation matrix can be shown to be

$$R = \begin{bmatrix} 0 & 0 & 0 & 0 & 0 \\ 0.2 & 0.2 & 0.2 & 0.1 & 0 \\ 0.5 & 0.5 & 0.4 & 0.1 & 0 \\ 0.8 & 0.8 & 0.4 & 0.1 & 0 \\ 1 & 0.8 & 0.4 & 0.1 & 0 \end{bmatrix} \tag{6.35}$$

Suppose the following question is posed: "Given the relation matrix equation (6.35), what is the change in $\{\textit{rudder angle}\}$ required if we have $\{\textit{yaw error about 15}\ ^\circ\}$?" where

$$\{\textit{yaw error about } 15^\circ\} = A' = \frac{0.1}{10.0} + \frac{0.6}{12.5} + \frac{1}{15.0} + \frac{0.6}{17.5} + \frac{0.1}{20.0} \tag{6.36}$$

Therefore we have

$$\begin{aligned} \min\left[\mu_{A'}(u), \mu_R(u,y)\right] &= \begin{bmatrix} \min[0,.1] & \min[0,.1] & \cdots & \min[0,.1] \\ \min[.2,.6] & \min[.2,.6] & \cdots & \min[0,.6] \\ \min[.5,1] & \min[.5,1] & \cdots & \min[0,1] \\ \min[.8,.6] & \min[.8,.6] & \cdots & \min[0,.6] \\ \min[1,.1] & \min[.8,.1] & \cdots & \min[0,.1] \end{bmatrix} \\ &= \begin{bmatrix} 0 & 0 & 0 & 0 & 0 \\ 0.2 & 0.2 & 0.2 & 0.1 & 0 \\ 0.5 & 0.5 & 0.4 & 0.1 & 0 \\ 0.6 & 0.6 & 0.4 & 0.1 & 0 \\ 0.1 & 0.1 & 0.1 & 0.1 & 0 \end{bmatrix} \end{aligned} \tag{6.37}$$

Hence from equation (6.31) we have

$$\begin{aligned} \mu_{B'}(y) &= \left[\max \begin{bmatrix} 0 \\ 0.2 \\ 0.5 \\ 0.6 \\ 0.1 \end{bmatrix} \max \begin{bmatrix} 0 \\ 0.2 \\ 0.5 \\ 0.6 \\ 0.1 \end{bmatrix} \cdots \max \begin{bmatrix} 0 \\ 0 \\ 0 \\ 0 \\ 0 \end{bmatrix} \right] \\ &= \begin{bmatrix} 0.6 & 0.6 & 0.4 & 0.1 & 0 \end{bmatrix} \end{aligned} \tag{6.38}$$

Therefore the answer to the question is:

$$\{change\ in\ rudder\ angle,\ \delta\} = \frac{0.6}{-10.0} + \frac{0.6}{-7.5} + \frac{0.4}{-5.0} + \frac{0.1}{-2.5} + \frac{0}{0} \quad (6.39)$$

Note that the resulting control set has a membership function whose values are significantly less than one. This implies that the knowledge of this situation is incomplete, but even under these conditions a control set can be realised.

Fuzzy Control Algorithms

A collection of control rules is, in Zadeh's terminology, a fuzzy algorithm and represents a powerful concept which can be used to provide fuzzy models and fuzzy controllers for use in control engineering. In this way, fuzzy set theory can be used to determine a collection of implication statements which causally link input and output fuzzy sets and thereby replace the need for the usual rigourous mathematical models. It should be noted that such relations are only possible if knowledge concerning the process under investigation is available *a priori*, although methods that incorporate a learning phase to generate the knowledge base are also becoming available (see Van der Rhee *et al.* [113]). Assuming that the relations can be formulated, a fuzzy control algorithm can be designed and implemented. Such algorithms are essentially a collection of "IF" and "THEN" statements which use the input/output fuzzy sets, and relations defined, to deduce the control signal necessary to achieve the desired response.

It is worth noting that although fuzzy controllers work with fuzzy concepts they have to accept non-fuzzy input variables (from system sensors), and produce non-fuzzy control outputs (to drive system actuators).

By way of example, consider a fuzzy algorithm relating yaw error, ε, and rudder control, movement, δ, consisting of two fuzzy conditional statements:

(a) IF { ε *much greater than* 10° } THEN { δ *greatly reduced* }

(b) IF { ε *about* 15° } THEN { δ *slightly reduced* }

where

$$\begin{aligned}
\{\varepsilon\ much\ greater\ than\ 10^\circ\} &= \frac{0}{10.0} + \frac{0.2}{12.5} + \frac{0.5}{15.0} + \frac{0.8}{17.5} + \frac{1}{20.0} \\
\{\varepsilon\ about\ 15^\circ\} &= \frac{0}{10.0} + \frac{0.6}{12.5} + \frac{1}{15.0} + \frac{0.6}{17.5} + \frac{0.1}{20.0} \\
\{\delta\ greatly\ reduced\} &= \frac{1}{-10.0} + \frac{0.8}{-7.5} + \frac{0.4}{-5.0} + \frac{0.1}{-2.5} + \frac{0}{0} \\
\{\delta\ slightly\ reduced\} &= \frac{0}{-10.0} + \frac{0.1}{-7.5} + \frac{0.5}{-5.0} + \frac{1}{-2.5} + \frac{0.6}{0}
\end{aligned}$$

The input of non-fuzzy variables presents no difficulty since these can be considered as special fuzzy sets (fuzzy singletons) which have only one member with a membership function value of 1. However, the production of non-fuzzy valued outputs from a fuzzy algorithm involves more work, as we now illustrate.

From equation (6.35) the relation matrix for rule (a) is

$$\begin{bmatrix} 0 & 0 & 0 & 0 & 0 \\ 0.2 & 0.2 & 0.2 & 0.1 & 0 \\ 0.5 & 0.5 & 0.4 & 0.1 & 0 \\ 0.8 & 0.8 & 0.4 & 0.1 & 0 \\ 1 & 0.8 & 0.4 & 0.1 & 0 \end{bmatrix} \quad \begin{matrix} \varepsilon^{\circ} \\ 10.0 \\ 12.5 \\ 15.0 \\ 17.5 \\ 20 \end{matrix} \tag{6.40}$$

and using equation (6.23) the relation matrix for rule (b) is

$$\begin{bmatrix} 0 & 0 & 0 & 0 & 0 \\ 0 & 0.1 & 0.5 & 0.6 & 0.6 \\ 0 & 0.1 & 0.5 & 1 & 0.6 \\ 0 & 0.1 & 0.5 & 0.6 & 0.6 \\ 0 & 0.1 & 0.1 & 0.1 & 0.1 \end{bmatrix} \quad \begin{matrix} \varepsilon^{\circ} \\ 10.0 \\ 12.5 \\ 15.0 \\ 17.5 \\ 20 \end{matrix} \tag{6.41}$$

Assuming the measured yaw error is 17.5° then this may be regarded as a yaw error "***much greater than*** 10°" and/or a yaw error "***about*** 15°".

If the former is accepted then the control set is given by

$$R^{a} = \frac{0.8}{-10.0} + \frac{0.8}{-7.5} + \frac{0.4}{-5.0} + \frac{0.1}{-2.5} + \frac{0}{0} \tag{6.42}$$

However, if the latter is accepted then

$$R^{b} = \frac{0}{-10.0} + \frac{0.1}{-7.5} + \frac{0.5}{-5.0} + \frac{0.6}{-2.5} + \frac{0.6}{0} \tag{6.43}$$

Several criteria can be utilised in the selection of a suitable control set. One may choose the rule which has the maximum membership value in its input set for the single-value measurement. In this case rule (a) would be selected. Alternatively, the rules of the algorithm can be considered as being connected by a logical ELSE which is a union operator (+ max) and therefore produces the following relation matrix:

$$R = R^{a} + R^{b} = \begin{bmatrix} 0 & 0 & 0 & 0 & 0 \\ 0.2 & 0.2 & 0.5 & 0.6 & 0.6 \\ 0.5 & 0.5 & 0.5 & 1 & 0.6 \\ 0.8 & 0.8 & 0.5 & 0.6 & 0.6 \\ 1 & 0.8 & 0.4 & 0.1 & 0.1 \end{bmatrix} \quad \begin{matrix} \varepsilon^{\circ} \\ 10.0 \\ 12.5 \\ 15.0 \\ 17.5 \\ 20 \end{matrix} \tag{6.44}$$

Hence in this example, the control set is

$$\frac{0.8}{-10.0} + \frac{0.8}{-7.5} + \frac{0.5}{-5.0} + \frac{0.6}{-2.5} + \frac{0.6}{0} \tag{6.45}$$

which reflects that both rules are significant for this single-valued measurement.

Several methods are available to determine a single control variable from the control output set. One method is to take the value which corresponds to the peak in the membership function, or averaging when there are several peaks. Another method is to form an average based on the shape of the membership function using a centre of area method (see Braae and Rutherford [17]). In general there is no clear reason for choosing one method in favour of another.

Several applications of fuzzy control algorithms have been demonstrated, see Mamdani and Assilian [76]; Kickert and van Nauta Lemke [66]; Rutherford and Carter [99]. The control area is receiving significant attention, and there is no doubt that rule-based applications will grow and improve with time. The same can also be said for the other forms of AI control systems that are beginning to emerge.

6.5 Parallel Processing Techniques

The dramatic developments in computer technology referred to earlier have mainly been achieved by improvements in manufacturing technology; relays in the first computers were replaced by vacuum tubes which in turn were replaced by transistors. Then came small scale integration, followed by medium, large and then very large scale integration (VLSI). With each transition, serial computer systems became faster and more powerful, allowing users to solve even more complex computational problems.

Unfortunately, this trend will soon come to an end because of a physical limitation, namely that signals cannot travel faster than the speed of light ($\approx 3 \times 10^8$ metres per second). To show how this hinders further improvements, assume that an electronic device can perform 10^{12} operations per second. It would then take longer for a signal to travel between two such devices one-half of a millimetre apart than it would take for either of them to process it! In other words, all the gains in speed obtained by building superfast electronic components are lost while one component is waiting to receive some input from another one. The problem cannot be solved by packing components closer and closer together since unwanted interactions can start to occur between the devices and reliability is reduced.

It appears that the only way around this difficulty is to use parallelism, so that several operations are performed simultaneously on different computing devices thereby reducing the overall processing time. The situation is synonymous with several people of comparable skills completing a job in a fraction of the time taken by one individual. Indeed, this fact has been widely recognised and significant interest in parallel computing techniques has been shown in the last ten years. It is expected that the "Fifth Generation" of computers will possess a high degree of parallelism (or concurrency) so that the rate of increase in computer system performances will be maintained. Such machines will no doubt have a strong impact on control engineering applications where large computing resources are required. The design of intelligent controllers will inevitably fall in this category where vast amounts of data need to be processed, and (possibly) complex reasoning conducted before the correct decision can be reached. Other areas where the real-time computational demands are too enormous to be adequately satisfied by sequential machines include the control of

(i) autonomous robots,
(ii) complex, distributed systems,
(iii) highly non-linear time variant multivariable systems, and
(iv) systems possessing very fast dynamics.

Some engineering investigations of parallel processing techniques have already begun (see IEE Computing and Control Colloquiums [47], [48]; IEE Electronics Colloquium [49]; Adey [1]; Freeman and Phillips [38]; Pritchard and Scott [96]), and this is likely to grow in importance. For this reason we provide an introduction to the area of parallel processing.

Parallel Computers

A *parallel computer* is defined as a computer with several processing units or processors, each of which can be used to process simultaneously different parts of the overall computational task. The results from each processor can be combined to produce an answer to the original problem. Such a philosophy is a radical change from the traditional sequential uniprocessor computer systems that form the majority of the systems in current use. The introduction of parallel computers poses several problems such as

(i) how to configure the architectures of the processors for application-optimised performance,
(ii) how to design the parallel algorithm for the architecture selected,
(iii) how to program the parallel computer for optimised performance while avoiding problems such as deadlock,
(iv) what are the programming languages to employ,
(v) how to ensure even loading of the processor paths, and
(vi) how to ensure that the communication overhead is minimised.

These issues are currently being investigated by computer scientists and other researchers and are beyond the scope of this text although we will touch on some of them in the following discussion. New titles and specialist groups are being set up to cater for the intensive research activity, see, for example, references in the *Journal of Parallel and Distributed Computing* [60]; *Concurrency: Practice and Experience* [21]; and *Parallel Computing* [92].

All computers, sequential or parallel, operate by executing instructions on some data. In parallel computers the structure of the streams of instructions (the algorithms) and data (the input to the algorithms) can be different for different architectures. Since many such computer networks can exist, it is convenient to classify them into the following four groups (see Sharp [102]):

(a) Single Instruction stream, Single Data stream (SISD);
(b) Multiple Instruction streams, Single Data stream (MISD);
(c) Single Instruction stream, Multiple Data streams (SIMD);
(d) Multiple Instruction streams, Multiple Data streams (MIMD).

(a) *SISD computers* are conventional computers consisting of one processor that receives a single stream of instructions and operates on a single stream of data, as shown in Figure 6.5.

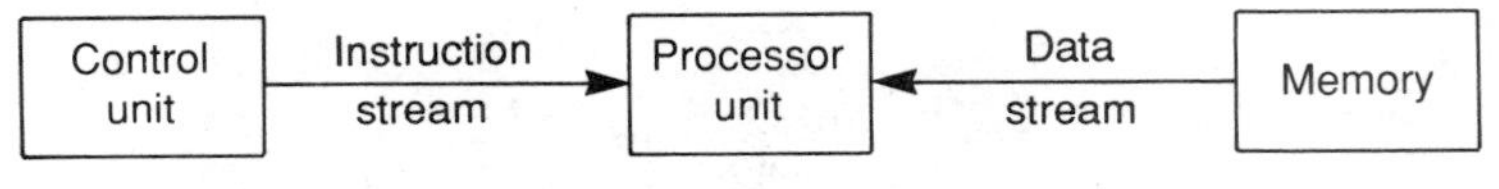

Figure 6.5 SISD computer

The majority of computers today are of this form, giving rise to the sequential Von Neumann bottleneck, see Stallings [104], and Hockney and Jesshope [45], in processing as all the instructions have to be executed by the single processor in a serial manner.

(b) *MISD computers* have several processors (say n) each with its own control unit, but having the same data stream as shown in Figure 6.6. Hence there are n instruction streams, one for each processor, and one data stream. Parallelism is achieved by letting the processors perform different operations at the same time on the same datum.

(c) *SIMD computers* consist of (say) n processors, all operated by a single control unit, but each with its own data stream, as shown in Figure 6.7. Hence each processor performs the same computations synchronously on different data. Communications between the processors are possible by having shared memory, or by having processors with local memory but interconnecting the processors in various configurations as shown in Figure 6.8, so that data and intermediate results can be exchanged. A combination of the two methods is also possible. Many

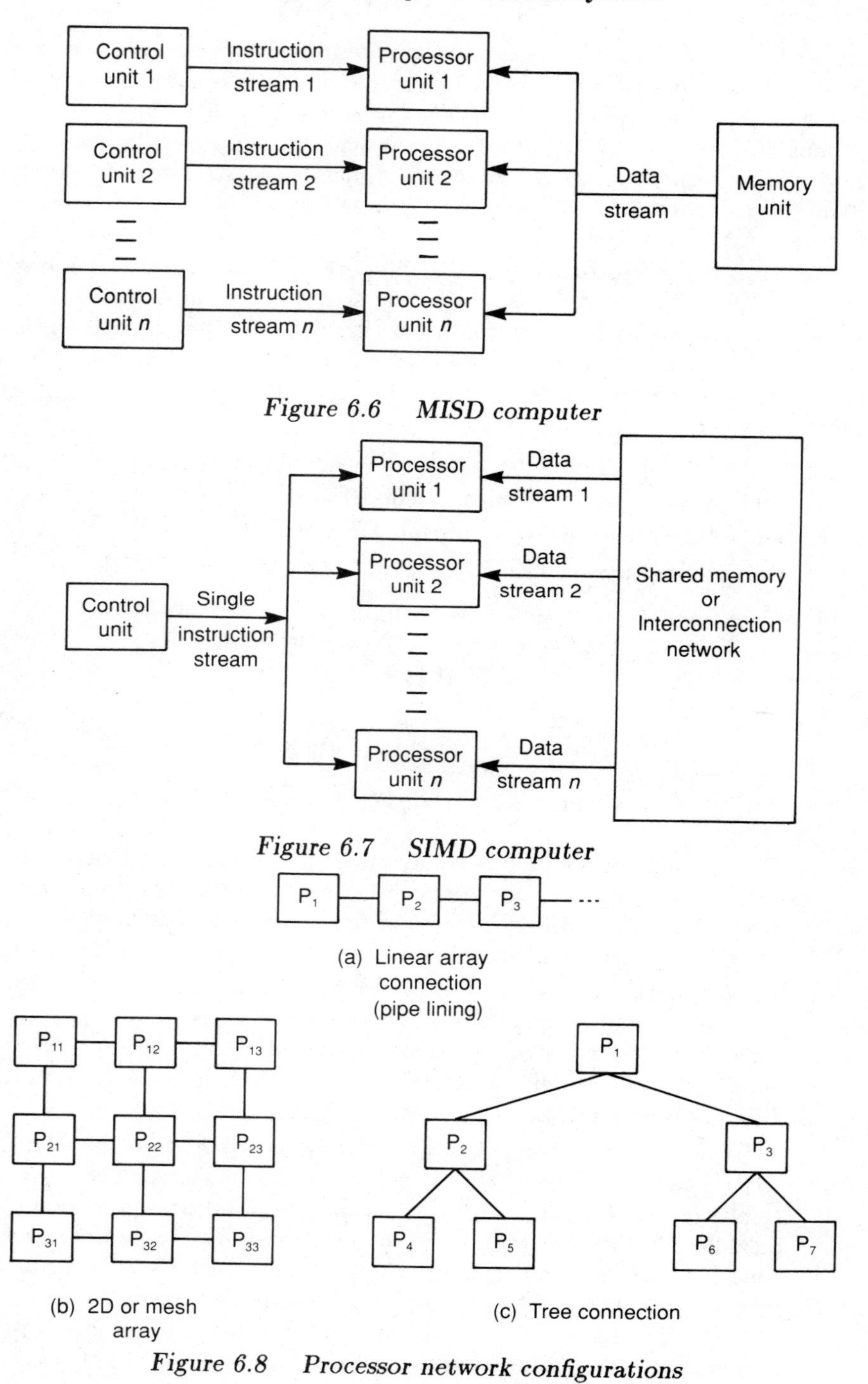

Figure 6.6 MISD computer

Figure 6.7 SIMD computer

Figure 6.8 Processor network configurations

other interconnection networks than those shown can be employed; these depend on the application, the number of processors and the precise computations required to be performed, see for example Akl [2], Decegama [26], and Bertsekas and Tsitsiklis [15].

(d) *MIMD computers* have (say) n processors, n streams of instructions and n streams of data as shown in Figure 6.9.

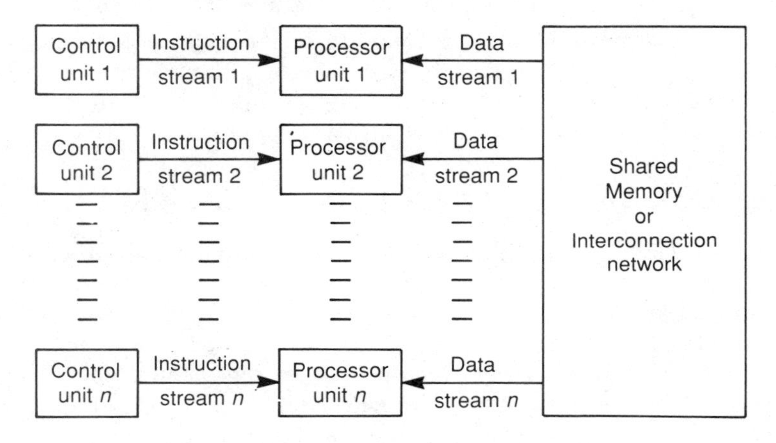

Figure 6.9 MIMD computer

These computers are the most general and most powerful class of parallel computers. The processors here are of the type used in MISD computers except that they have their own control units which issue different instructions for each processor. Hence the processors are potentially all executing different programs on different data while solving different subproblems of an overall problem. The technique is also known as functional decomposition since each processor node potentially performs a different function and so operates in an asynchronous fashion. As with SIMD computers, communication between processors can be performed through a shared memory, or an interconnection network, or a mixture of the two.

Shared memory systems can give rise to memory contention problems, where adding extra processors does not give improved performance because processors have to wait (idle) while access to memory is being achieved. An interconnected processor system where each processor has its own local memory, does not suffer from this problem, and adding extra processors gives further improvements provided the communication overheads are not excessively increased.

Inmos Ltd have recently developed the transputer (see Inmos [54], [55], [56]) which allows MIMD computer architectures to be implemented relatively easily. The transputer is in fact a generic name for a family of VLSI devices each with a CPU, on-chip memory, an external memory interface and communication links for direct point to point connection to other transputers. Parallel computer systems can thus be constructed from a network of interconnected transputers which operate simultaneously and communicate through their links. A precise programming language, "occam", has been developed by Inmos specially for the transputer, to provide parallel constructs and communication synchronization.

By way of introduction it is better to keep the discussion general at this stage and we will not go into details concerning the transputer and occam. Further discussion on these aspects is given in appendix C. Here we continue the discussion on other aspects of MIMD computers.

Algorithms for Parallel (MIMD) Computers

Since MIMD computers involve a number of interconnected processors, solving interrelated subproblems in an asynchronous manner, the task of programming algorithms for these machines is rather complicated. To appreciate the level of difficulty, it is important to distinguish between the notion of a process and that of a processor. An asynchronous algorithm is a collection of processes, some or all of which are executed simultaneously on a number of processors. The mode of operation is as follows.

(a) The parallel algorithm starts its execution on a selected processor, which in turn may create a number of computational tasks, or processes to be performed. A process therefore corresponds to a section of the algorithm and there may be several processes with the same algorithm section, each with a different parameter.

(b) Once a process is created, it must be executed on a processor. If a free processor is available, the process is assigned to the processor which then performs the computations specified by the process. On the other hand, if no processor is available the process is queued and waits for a processor to become free.

(c) When a processor completes execution of a process, it becomes free and is assigned another process, if there happens to be one waiting in the queue. If not the processor is queued, and waits for a process to be created.

The order in which the processes are executed by processors is selectable – for example, first-in-first-out or last-in-first-out policies can be easily implemented. Also, the availability of a processor is sometimes not sufficient for the processor to be assigned a waiting process. An additional condition, such as the arrival of additional data, may have to be satisfied before the process starts.

Similarly, if a processor has already been assigned a process and an unsatisfied condition is encountered during execution, then the processor is freed. When the condition for resumption of that process is later satisfied, a processor (not necessarily the original one) is assigned to it. These are some of the scheduling problems that characterise the programming of MIMD computers. They need to be solved efficiently if parallel computers are to become really useful. Note that the programming requirements of SIMD computers are significantly easier because of their synchronous nature.

Programming Parallel Computers

There are now many languages that support parallel programming; Parallel-Pascal, Modula 2, Parallel-C, occam, Parallel-Fortran, Ada, etc. There is little point in looking at all languages with concurrent programming facilities, but what is important is to be aware of, and to appreciate, the important features necessary for a good parallel programming language. An outline of how some of these are provided is therefore given. As the following discussion will show, several interrelated issues unique to parallel computing need to be considered – these include the following.

Method of communication

The common methods are message passing, unprotected shared (global) variables, shared data protected by modules (Modula 2 – see Bennett [13]) and the rendezvous (Ada see Young [119], and Gehani and McGettrick [39]).

Process synchronization

The parallel processes need to be synchronised so that they can enforce sequencing restrictions on each other. The methods employed include signals, synchronised send (see Hoare [43]), buffers, path expressions, events, conditions, queues and rendezvous.

Process creation

Concurrent processes are either all created statically at the beginning of program execution, or they are each initiated dynamically during execution. Both these cases need to be accommodated adequately.

Process topology

The interconnection structure can be

- static, that is, the communication links between the processes either do not change during execution, or if they do, the dynamic changes occur in a way that is predictable at compile-time. Hence the manner in which processes are added (or deleted) dynamically, are known before execution of the program.

- dynamic, that is, the links may change during execution in a way that cannot be known prior to run-time.

Separate compilation

The language provides the ability to compile a coded module without having all the code for the rest of the program.

Real-time support

A language can include features that make real-time programming possible (see chapter 5). Explicit programmer control over the order of process execution is essential as well as the concept of time-out in process communications. A process that sends a message and awaits a reply must assume a catastrophic occurrence, and take appropriate action, if the desired response is not received within a set time (a real-time clock is necessary for this). If such time-out problems are not accounted for in the design stages, parallel computers are open to the possibility of process deadlock (a process waiting for ever for a response that never comes). Other important aspects are the ability to communicate directly with external hardware devices, and the need for a well structured interrupt facility.

Performance of Parallel Computers

Since the speeding up of computations is one of the main reasons for building parallel computers, it is important to evaluate the running time of a parallel algorithm. This is defined to be the time elapsed from the moment the first process starts on the first processor, to the moment the last processor ends its computing. Then the speedup of a parallel algorithm is defined as

$$\text{Speedup} = \frac{T_{seq}}{T_{par}} \tag{6.46}$$

where T_{seq} is the worst-case running time for the fastest sequential algorithm problem, and T_{par} is the worst-case running time for the parallel algorithm. Clearly the larger the speedup, the better the parallel algorithm.

Another aspect that can be used to categorise the performance of parallel systems is to calculate the efficiency of running the parallel algorithm, where

$$\text{Efficiency} = \frac{\text{sequential processing time}}{\text{parallel running time} \times \text{number of processors}} \tag{6.47}$$

If a single processor requires T seconds to execute an algorithm then, theoretically at least, it should take two processors working together $T/2$ seconds; on an n processor system it should take T/n to execute the parallel algorithm, giving an ideal efficiency of unity. This theoretical limit is, generally, not achievable in practice because the sequential algorithm may not be suitable for parallelisation, and parallelisation may in fact result in a degradation in performance. However the closer the efficiency is to unity, the better is the implementation on the parallel computer hardware. In addition, the multiple processors, in general, require communications with other processors for the overall solution to be computed. Such communications can account for a sizeable fraction of the total time, as processors may have to idle if the communicating processor is not ready. A balance between using few processors with little inter-communications, and many processors with large communication overheads, has to be reached. This is commonly referred to as the granularity of parallelisation. For further discussion on these and other matters for parallel systems see Bertsekas and Tsitsiklis [15], and Akl [2].

6.6 Parallel Processing in Control Engineering

In this section we look at introducing parallel processing techniques into control engineering applications so that improved performance levels and faster cycle times are achievable, as well as offering greater potential in reliability as a result of the fault tolerance aspects encompassed by multiprocessor systems. As discussed earlier in section 6.5, there are many different forms of parallel computers; some of these have many simple processors connected in fixed architectures, whereas others have more powerful processors and are reconfigurable (see Hockney and Jesshope [45]). In control applications the computations required to be performed can vary tremendously in terms of processing complexity, and sequential to parallel transformation efficiency. To maximise the effectiveness of parallel processing methods in such varied cases it is imperative that the control engineer has the ability to design the best computer architecture for his application. In this way he can subdivide the overall control computational task into sub-tasks which can be mapped onto tailor-made processor networks for optimised performance. A fixed architecture multiprocessor system may be suitable for some applications but, in general, cannot offer the flexibility required, and so such computer implementations will be inefficient in terms of communication overheads and processor idle times. Therefore in implementing parallel processing methods the primary objective is to distribute the overall computational task into a number of co-operating processors in such a way that all the processors are fully utilised and close to linear speedup is achieved.

Considering a typical control task and how it can be solved using a parallel computer system, it is necessary to divide the overall problem into its smaller constituents. For example, in a computer control application where clearly the overall task is to control the system under consideration by using the computer. This can be viewed as being composed of smaller sub-tasks such as:

(i) interface system to computer for data transfer,
(ii) modelling of system,
(iii) validation of model,
(iv) controller design and implementation,
(v) fault detection and remedial action,
(vi) performance assessment and data logging, and
(vii) user interface (usually graphics).

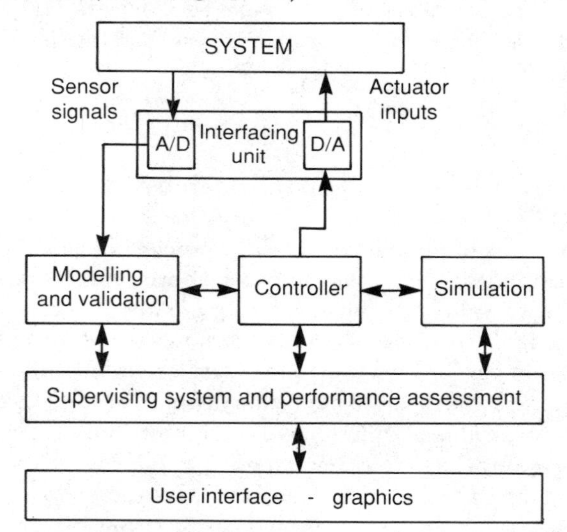

Figure 6.10 Functional decomposition of a general control problem

Each of these sub-tasks can be considered as a component in the overall computational task which needs to be executed for effective control. In this way the component sub-tasks can be mapped onto a computer system structure as shown in Figure 6.10. Here each block can be individually designed to contain a network of processors necessary for solving the sub-task within specified time requirements. Parallel processing systems based on transputer hardware allows such a functional decomposition of the control problem for solving computational problems in a straightforward manner, and an active control application example is presented in appendix C.

Having broken down the control task into smaller sub-tasks, the procedures for execution need to be analysed to assess their suitability for parallelisation. Since each application is different with different characteristics, it is difficult to establish universal parallelisation procedures which will apply to all instances, but some of the sub-tasks are generic and so discussion on these is worthwhile. Two of the most commonly occurring operations in control applications are matrix algebra and numerical integration. Sample methods for how these, up to now, sequential procedures can be parallelised follows next.

Matrix Algebra

Matrix operations are ideal candidates for execution on multi-processor systems and obvious parallelisation of the sequential loops can be employed. For example, the inner product of n-dimensional vectors x and y, given by

$$x \cdot y = \sum_{i=0}^{n} x_i y_i \tag{6.48}$$

can be computed using n processors with each processor i computing the product $x_i y_i$ and then adding the $x_i y_i$ to form $x \cdot y$.

Matrix operations such as addition, multiplication and transposition can be parallelised by distributing the arithmetic operations over several processors. Care needs to be taken so that the numerical calculations are not distributed too sparsely over a large number of processors so that excessive communications are avoided.

As an example, consider the $n \times n$ Riccati matrix equation that arises in LQP continuous optimal control problems, see Owens [90]:

$$-\dot{P}(t) = Q + A^T P(t) + P(t) A - P(t) BR^{-1} B^T P(t) \tag{6.49}$$

This can be solved, in principle, on any number of processors connected in a pipeline, where each processor used computes different elements of the $\dot{P}(t)$ matrix. For instance if $n = 50$ and 10 processors are used, processor 1 can be used to compute rows 1 to 5 of $\dot{P}(t)$, processor 2 for rows 6 to 10, etc. However to form the matrix product terms such as $A^T P(t)$, it is necessary to communicate the elements that are not held locally. For example, to form row 4 of $A^T P(t)$, we need to form the matrix product

$$\{row\ 4\ of\ A^T P(t)\} = \{row\ 4\ of\ A^T\} \times \{matrix\ P(t)\} \tag{6.50}$$

For matrices of large dimension, such communication overheads are small compared with the total computation time of each processor. If, however, the matrices are of small dimension then the communication time can form a large proportion of the overall time and thus seriously impair the performance.

In these smaller dimensional cases it may be better to use an alternative approach where the matrices are not distributed but the equation is split into its component terms. In this way different processors can be used to compute the three different terms on the right-hand side of equation (6.49), where processor 1 computes the matrix product $-P(t)\,BR^{-1}B^T P(t)$, processor 2 computes $P(t)\,A$, and processor 3 computes $A^T P(t)$. The result from these three worker processors can be fed in parallel to a fourth master processor where the $\dot{P}(t)$ matrix is formed. Such a strategy requires no inter-processor communication (except that between the master and the workers), since a tree type structure can be employed connecting the three workers to the master. For equations involving 10×10 matrices this method has been shown to be over six times faster than a ten processor pipeline configuration approach when the Riccati equation is integrated, see Kourmoulis [68]. Hence good results can only be achieved if some form of optimised parallelisation procedure is used, and so it is important to give due consideration to the sequential problem to be tackled and how the solution implementation can be best parallelised.

Numerical Integration

In most control applications it is a requirement that differential equations be solved so that the system/model can be simulated. For large dimensional systems with fast dynamics, or for stiff systems, the processing required for the numerical integration can place a heavy computational burden, and so it is an obvious operation to execute on a parallel computer system. Several schemes have been suggested for parallelising numerical integration procedures, and some of these have been discussed in Franklin [37]. The majority of the methods modify a sequential method in some way so that multiple processing paths are introduced. To assist our discussion, consider the set of n non-linear ordinary differential equation to be solved as

$$\dot{x}(t) = f(x(t), u(t), t) \tag{6.51}$$

An obvious approach is to divide the n equations into smaller groups which can be distributed over the available processors. Since the function evaluations are then done in parallel, a good deal of time can be saved over a single processor solution. However for optimal performance, the processors have to be loaded evenly so that none are left idling for long times. In addition the coupling between the $\dot{x}_i$ equations will have a strong effect on the performance since this will determine the level of communications between the processors.

The Adams–Bashforth predictor–corrector formulae have been modified by Miranker and Liniger [85] to permit parallel execution. The traditional second-order predictor–corrector formulae for solving the equations

(6.51) are

$$x_{n+1}^p = x_n^c + \frac{\Delta t}{2}\left\{3f_n^c - f_{n-1}^c\right\} \quad (6.52)$$

$$x_{n+1}^c = x_n^c + \frac{\Delta t}{2}\left\{3f_{n+1}^p + f_n^c\right\} \quad (6.53)$$

where

x_n = an approximation to x at time t_n
x_n^p = predicted value of x_n
x_n^c = corrected value of x_n
f_n^p = $f\left(x_n^p, u_n, t_n\right)$
f_n^c = $f\left(x_n^c, u_n, t_n\right)$
Δt = time step

and the control u is a known function of time. This method is a strictly sequential procedure. Assuming that we are at time t_n then the procedure is as follows:

(i) x_{n+1}^p is computed using equation (6.52);
(ii) having computed x_{n+1}^p allows f_{n+1}^p to be computed;
(iii) x_{n+1}^c is then computed using equation (6.53);
(iv) having computed x_{n+1}^c, f_{n+1}^c can be computed;
(v) update $t_n = t_{n+1}$ and repeat procedure.

The parallel form for equations (6.52) and (6.53), suggested by Miranker and Liniger, is

$$x_{n+1}^p = x_{n-1}^c + 2\Delta t f_n^p \quad (6.54)$$

$$x_n^c = x_{n-1}^c + \frac{\Delta t}{2}\left\{f_n^p - f_{n-1}^c\right\} \quad (6.55)$$

where the predictor and corrector sequences of computation are separated, and hence can be executed simultaneously on two processors with each processor requiring intermediate data calculated by the other processor. Although inter-processor communication is required, the method has been shown to work well since the two equations for the two processors involve approximately the same amount of calculation, see Jones [58], [59]. It is possible to extend the method to a greater number of processors by using higher order formulae. Krosel and Milner [67] show that good accuracy and speedup is obtained with 8 processors.

In most cases when a sequential algorithm is parallelised in this way, a degree of degradation occurs in numerical stability. The reason lies in the fact that previous values of the variables are used in the different parallel paths, in contrast to the most recent calculated values used in the sequential solution implementation. Care must be taken to ensure that stability and good accuracy are maintained when attempting to speed up numerical integration in this way (see Miranker and Liniger [85]).

Fault Tolerance

Another feature offered by using parallel computers in control applications is that of fault tolerance. In sequential computer controlled methods, if the computer develops a fault the control function is lost unless a secondary standby computer takes over. In parallel computer systems, it is possible to distribute the computation in such a way that failure of one processor can be tolerated by reconfiguring the computational task over the operational processors. In this way, highly reliable systems can be obtained without resorting to purely redundant systems. Inevitably in any fault tolerant system, a degree of redundancy is necessary, but multi-processor based implementations offer the potential of reducing the hardware duplication while maintaining high reliability levels.

In addition, parallel computer systems permit an alternative approach in the design procedure. Since the overall computational task is distributed over several processors, if one device fails, the systems can be designed so that the control function is not totally lost but the performance is degraded since the remaining processors can be reconfigured to execute the tasks. The concept of graceful degradation is often used to describe the "smooth" decline in the control performance as more and more processors fail. With a small level of redundancy the performance levels can be maintained by allocating the tasks of the failed processor to standby devices.

We illustrate a parallel fault tolerant control system design by considering the following example, see Ayuk [7] for further details. Assume we wish to control the third-order, two input and single output system represented by

$$\begin{bmatrix} \dot{x}_1(t) \\ \dot{x}_2(t) \\ \dot{x}_3(t) \end{bmatrix} = \begin{bmatrix} -2 & 1 & 0 \\ 2 & -1 & 0 \\ 0 & 0 & 1 \end{bmatrix} \begin{bmatrix} x_1(t) \\ x_2(t) \\ x_3(t) \end{bmatrix} + \begin{bmatrix} 1 & 0 \\ 0 & 1 \\ 1 & 1 \end{bmatrix} \begin{bmatrix} u_1(t) \\ u_2(t) \end{bmatrix} \tag{6.56}$$

$$y(t) = \begin{bmatrix} 0 & 0 & 1 \end{bmatrix} x(t) \tag{6.57}$$

The open-loop poles of the system can be shown to be located at $s = 0, 1$ and -3. Hence the system is unstable because of the right-half plane pole at $s = 1$, and some form of compensation is required. We propose to use state-feedback to modify the system matrix from A to $(A + BF)$ as discussed in chapter 4.

The system is seen to be controllable with respect to both inputs, and so state-feedback can be applied to either u_1, or u_2, or both u_1 and u_2 together. To introduce a degree of fault tolerance it is better to design feedback vectors for u_1 and u_2 separately and use only one control in practice. If this control fails it is possible to switch to the other control and retain control.

Performing the feedback vector designs is straightforward, as discussed in chapter 4. Assuming that the required closed-loop characteristic equation is

$$s\,(s+3+j4)\,(s+3-j4) = s^3 + 6s^2 + 25s = 0 \qquad (6.58)$$

then the feedback vectors are found to be

(i) feedback on control u_1 only, $f_1 = [f_{11} \quad f_{12} \quad f_{13}]$ where

$$f_1 = [4 \quad -2 \quad -8] \qquad (6.59)$$

(ii) feedback on control u_2 only, $f_2 = [f_{21} \quad f_{22} \quad f_{23}]$ where

$$f_2 = [-8 \quad 4 \quad -8] \qquad (6.60)$$

A fault tolerant version of this design can be implemented on a parallel computer system as shown in Figure 6.11.

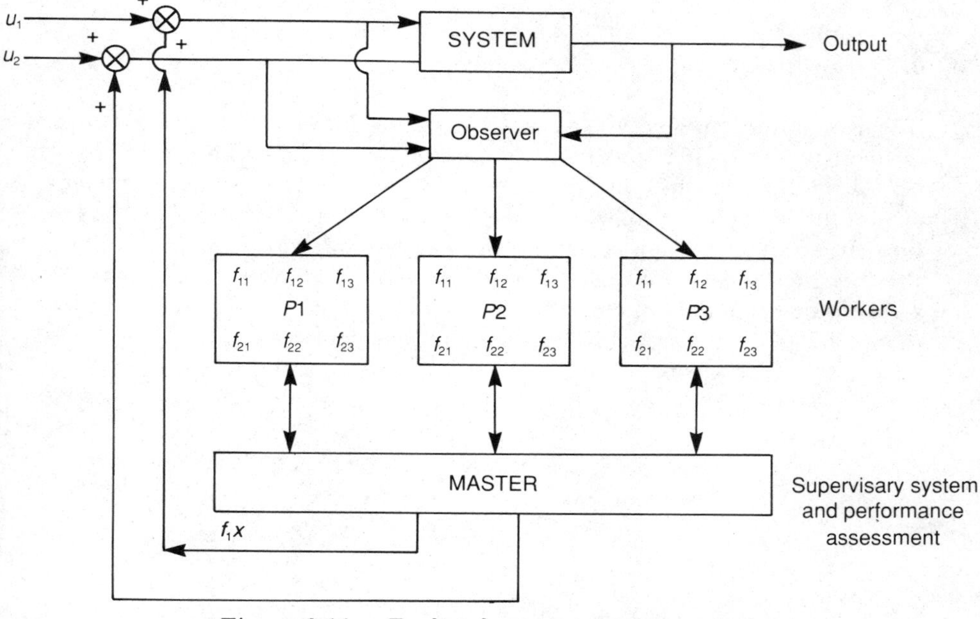

Figure 6.11 Fault tolerant computer control

The three workers have all the states and feedback gains available to them. Under normal conditions it is possible to allocate processor $P1$ to handle state x_1, processor $P2$ to state x_2, etc., and only have state-feedback active on one control loop, say u_1. Hence $P1$ receives the complete state vector $x\,(t)$ from the observer processor block, multiplies x_1 by the gain f_{11} and transmits the result to the supervisory system and performance assessment block (the *Master* block). $P2$ does the same for x_2 and $P3$ for x_3. The three numbers are added in the *Master* block to give the feedback signal $(= f_{11}x_1 + f_{12}x_2 + f_{13}x_3)$ as required.

Under faulty conditions the supervisory system can instruct the workers $P1 - P3$ accordingly so that the control is retained. For example, if the input u_1 fails then the workers have to be directed to use the feedback vector f_2 and apply the calculated feedback signal to the input u_2 loop.

If a worker fails, the other functional processors have to handle the state of the failed processor, etc. Other faults such as communication failures, incorrect signals and software calculation errors can also be handled. The fault condition first needs to be detected (there are standard procedures for this, see Ayuk [7]), and then the computing architecture reconfigured to retain the control function.

Another commonly used approach is to use redundant hardware systems to give rise to a voting scheme so that the faults can be masked out. For further discussion on fault tolerance control systems, see Ayuk [7].

6.7 Summary

In this chapter we have looked at some of the latest trends in computer technology and their implications to control applications. The areas of artificial intelligence and parallel processing together provide an extremely powerful technique for improving the computer control systems of the future. It is ironic to think that these systems are likely to be able to mimic the thought process and operate as humans do. The purely mathematical algorithmic techniques of current systems will surely be replaced in part by these "vaguely thinking" superfast computer systems.

Bibliography and References

[1] Adey R A (Editor), Parallel Processing in Engineering Applications, Springer-Verlag, Berlin, 1990.

[2] Akl S G, The Design and Analysis of Parallel Algorithms, Prentice-Hall, 1989.

[3] Alty J L and Coombs M J, Expert Systems: Concepts and Examples, NCC Publications, 1984.

[4] Artificial Intelligence: An International Journal, Special Volume on Qualitative Reasoning about Physical Systems, Vol 24, Numbers 1-3, Dec 1984.

[5] Astrom K J and Wittenmark B, Computer Controlled Systems: Theory and Design, Prentice-Hall, New Jersey, USA, 1984.

[6] Astrom K J, Anton J J and Arzen K E, Expert Control, Automatica, Vol 22, (3), pp 277-286, 1986.

[7] Ayuk J T, Fault tolerant reconfigurable architectures for flight control, Masters Thesis, University of Sheffield, 1989.

[8] Balas M J, Feedback control of flexible systems, IEEE Transactions on Automatic Control, Vol. AC-23, no. 4, pp. 673-679, 1978.

[9] Balas M J, Enhanced modal control of flexible structures via innovations feedthrough, International Journal of Control, Vol. 32, no. 6, pp. 983-1003, 1980.

[10] Banks S P, Control Systems Engineering, Prentice-Hall, 1986.

[11] Barney G C, Intelligent Instrumentation: Microprocessor Applications in Measurement and Control, Prentice-Hall, 1985.

[12] Barr A and Feigenbaum E A, The Handbook of Artificial Intelligence, Vol 1, Pitman Books Ltd, 1981.

[13] Bennett S, Real-Time Control Systems, Prentice-Hall, 1987.

[14] Berg M C, Amit N and Powell J D, Multi-rate Digital Control System Design, IEEE Trans Aut Control, Vol 33, (12), Dec, pp 1139-1150, 1988.

[15] Bertsekas D P and Tsitsiklis J N, Parallel and Distributed Computation: Numerical Methods, Prentice-Hall, 1989.

[16] Boykin W H and Frazier B D, Multi-rate Sampled-Data Systems Analysis via Vector Operators, IEEE Trans Aut Control, pp 548-551, 1975.

[17] Braae M, and Rutherford D A, Fuzzy relations in a control setting, Tech report Mo 359, control Systems Centre, UMIST, 1977.

[18] Bryson A E and Ho Y C, Applied Optimal Control, Blaisdell, 1969.

[19] Chen C-T, Linear System Theory and Design, Holt, Rinehart and Winston, New York, 1984.

[20] CODAS Operating Manual, Golten and Verwer Parteners, 33 Moseley Rd, Cheadle Hulme, Cheshire.

[21] Concurrency: Practice and Experience, ISSN 1040-3108, John Wiley & Sons Ltd, New York, USA.

[22] Coon G A, How to Find Controller Settings from Process Characteristics, Control Eng, Vol 3 (5), p 66, 1956.

[23] Coon G A, How to Set Three Term Controllers, Control Eng, Vol 3 (6), p 21, 1956.

[24] Davis R, Buchanan B G and Shortliffe E H, Production Systems as a Representation for a Knowledge-Based Consultation Program, Artificial Intelligence, Vol 8, (1), pp 15-45, 1977.

[25] D'Azzo J J and Houpis C H, Linear Control System Analysis and Design: Conventional and Modern, McGraw-Hill, Tokyo, 1981.

[26] Decegama A L, The Technology of Parallel Processing, Vol 1, Prentice-Hall, 1989.

[27] Denham M J and Laub A J (Editors), Advanced Computing Concepts and Techniques in Control Engineering, Springer-Verlag, 1988.

[28] Di Stefano III J J, Stubberud A R and Williams I J, Feedback and Control Systems, Schaum's Outline Series, McGraw-Hill, 1967.

[29] Dorf R C, Modern Control Systems, Addison-Wesley, 1986.

[30] Duda R O, Gaschnig J G and Hart P E, Model Design in the PROSPECTOR Consultation System for Mineral Exploitation, in Expert Systems in the Micro-Electronic Age, Edited by D Mitchie, Edinburgh University Press, pp 153-167, 1979.

[31] Elgerd O I, Control Systems Theory, McGraw-Hill, 1967.

[32] Fleming W H and Rishel R W, Deterministic and Stochastic Control, Springer-Verlag, New York, 1975.

[33] Flowers D C and Hammond J L, Simplification of the Characteristic Equation of Multi-rate Sampled Data Control Systems, IEEE Trans Aut Control, pp 249-251, 1972.

[34] Forgyth R (Editor), Expert Systems: Principles and Case Studies, Chapman and Hall, 1984.

[35] Franklin G F, Powell J D and Workman M L, Digital Control of Dynamic Systems, 2nd edition, Addison-Wesley, New York, 1990.

[36] Franklin G F, Powell J D and Emami-Naeini A, Feedback Control of Dynamic Systems, Addison-Wesley, New York, 1986.

[37] Franklin M A, Parallel solution of ordinary differential equations, IEEE Trans Computers, C-27 (5), pp 413-430, 1978.

[38] Freeman L and Phillips C (Editors), Applications of Transputers 1, IOS Press, Amsterdam, 1989.

[39] Gehani N and McGettrick A D (Editors), Concurrent Programming, Addison-Wesley, 1988.

[40] Goodwin G C and Sin K S, Adaptive Filtering, Prediction and Control, Prentice-Hall, 1984.

[41] Hayes-Roth F, Waterman D and Lenat D (Editors), Building Expert Systems, Addison-Wesley, 1983.

[42] Hestenes M R, Calculus of Variations and Optimal Control Theory, John Wiley, 1966.

[43] Hoare C A R, Communicating Sequential Processes, Communications of the ACM, 21(8), pp 667-677, 1978.

[44] Hoare C A R and Perrott R H, Operating Systems Techniques, Academic Press, 1972.

[45] Hockney R W and Jesshope C R, Parallel Computers 2, Adam Hilger, Bristol, 1988.

[46] Houpis C H and Lamont G B, Digital Control Systems, Theory, Hardware, Software, McGraw-Hill, New York, 1985.

[47] IEE Computing and Control Colloquium on, The Transputer: Applications and Case Studies, Digest No 1986/91, 1986.

[48] IEE Computing and Control Colloquium on, Recent Advances in Parallel Processing for Control, Digest No 1988/94, 1988.

[49] IEE Electronics Colloquium on, Transputers for Image Processing Applications, Digest No 1989/22, 1989.

[50] Inman D J, Vibration with control measurement and stability, Prentice-Hall, NJ, 1989.

[51] Inmos Ltd, Communicating Process Architecture, Prentice-Hall, 1988.

[52] Inmos Ltd, occam 2 Reference Manual, Prentice-Hall, 1988.

[53] Inmos Ltd, Transputer Development System, Prentice-Hall, 1988.

[54] Inmos Ltd, Transputer Instruction Set, Prentice-Hall,1988.

[55] Inmos Ltd, Transputer Reference Manual, Prentice-Hall, 1988.

[56] Inmos Ltd, Transputer Technical Notes, Prentice-Hall, 1989.

[57] Jackson P, Introduction to Expert Systems, Addison-Wesley, 1986.

[58] Jones D I, Parallel processing for computer aided design of control systems, Research report, University College of north Wales, 1985.

[59] Jones D I, Occam structures in control, IEE Workshop on parallel processing and control - the transputer and other architectures, Digest 1988/95, 1988.

[60] Journal of Parallel and Distributed Computing, ISSN 0743-7315, Academic Press Inc, San Diego, USA.

[61] Jury E I, Theory and Application of the z-Transform Method, John Wiley, 1964.

[62] Jury E I, A Note on Multi-rate Sampled Data Systems, IEEE Trans Aut Control, pp 319-320, 1967.

[63] Jury E I and Blanchard J, A Stationary Test for Linear Discrete Systems in Table Form, Proc IRE, Vol 49, pp 1947-1948, 1961.

[64] Katz P, Digital Control Using Microprocessors, Prentice-Hall, London, 1981.

[65] Kerridge J, occam Programming: A Practical Approach, Blackwell Scientific Publications, Oxford, 1987.

[66] Kickert W J M and van Nauta Lemke H R, Application of a Fuzzy Controller in a Warm Water Plant, Automatica, 12, pp 301-308, 1976.

[67] Krosel S M and Milner E J, Application of integration algorithms in a parallel processing environment for the simulation of jet engines, Proc IEEE Annual Simulation Symposium, pp 121-143, 1982.

[68] Kourmoulis P K, Parallel processing in the simulation and control of flexible beam structure systems, Ph D Thesis, University of Sheffield, 1990.

[69] Kourmoulis P K and Virk G S, Parallel processing in the simulation of flexible structures, IEE Colloquium on recent advances in parallel processing for control, Digest No 1988/94, 1988.

[70] Kuo B C, Digital Control Systems, Holt-Saunders, Tokyo, 1980.

[71] Kuo B C, Automatic Control Systems, Prentice-Hall, New Jersey, 1987.

[72] Lapidus L and Seinfeld J H, Numerical Solution of Ordinary Differential Equations, Academic Press, 1971.

[73] Lee E B and Markus L, Foundations of Optimal Control Theory, John Wiley, 1967.

[74] Leigh J R, Applied Digital Control (Theory, Design and Implementation), Prentice-Hall, 1985.

[75] Leitch R, Qualitative Modelling of Physical Systems for Knowledge Based Control, in Advanced Computing Concepts and Techniques in Control Engineering, edited by M J Denham and A J Laub, NATO ASI Series, Springer-Verlag, 1988.

[76] Mamdani E H and Assilian S, A Fuzzy Logic Controller for a Dynamic Plant, Int J Man-Machine Studies, 7, pp 1-13, 1975.

[77] Manson G A, The Granularity of Parallelism required in Multi-Processor Systems, in SERC Vacation School on Computer Control, IMC London, 1987.

[78] McDermott J, R1: An Expert in the Computer Systems Domain, Proc of the AAAI-80, 1, pp 269-271, 1980.

[79] McDermott J, R1: A Rule-Based Configurer of Computer Systems, Artificial Intelligence, 19, pp 39-88, 1982.

[80] Meirovitch L, Elements of Vibration Analysis, McGraw-Hill, New York, 1986.

[81] Meirovitch L and Öz H, Computational aspects of the control of large flexible structures, Proceedings of the 18th IEEE Conference on Decision and Control, Fort Launderdale, pp. 220-229, 1979.

[82] Meirovitch L and Öz H, Modal-space control of large flexible spacecraft possessing ignorable coordinates, Journal of Guidance and Control, Vol. 3, no. 6, pp. 569-577, 1980.

[83] Mellichamp D A (Editor), Real-Time Computing, Van Nostrand Reinhold Co, New York, 1983.

[84] Michie D (Editor), Introductory Readings in Expert Systems, Gordon and Breach Science Publishers, 1982.

[85] Miranker W L and Liniger W M, Parallel methods for the numerical integration of ordinary differential equations, Math Comp, 21, pp 303-320, 1967.

[86] Newland D E, Mechanical vibration: Analysis and computation, John Wiley & Sons, New York, 1989.

[87] Norman D, Some Observations on Mental Models, in Mental Models, edited by D Gentner and L A Stevens, Lawerence Earlbaum Associates, 1983.

[88] Ogata K, Modern Control Engineering, Prentice-Hall, New Jersey, 1970.

[89] Ogata K, Discrete-Time Control Systems, Prentice-Hall, New Jersey, 1987.

[90] Owens D H, Multivariable and Optimal Systems, Academic Press, 1981.

[91] Papoulis A, Probability, Random Variables, and Stochastic Processes, McGraw-Hill, 1965.

[92] Parallel Computing, ISSN 0167-8191, North Holland, Amsterdam, Netherlands.

[93] PC-MATLAB User Guide, The MathWorks Inc, 20 North Main Street, Suite 250, Sherborn, MA 01770, USA.

[94] Phillips C L and Harbor R D, Feedback Control Systems, Prentice-Hall, New Jersey, 1988.

[95] Pontryagin E F, Boltanskii V G, Gamkrelidze R V and Mishchenko [E F,] The Mathematical Theory of Optimal Processes, Pergamon Press, 1964.

[96] Pritchard D J and Scott C J (Editors), Applications of Transputers 2, IOS Press, Amsterdam, 1990.

[97] Raible R H, A Simplification of Jury's Tabular Form, IEE Trans on Automatic Control, Vol AC-19, pp 248-250, 1974.

[98] Raven F H, Automatic Control Engineering, McGraw-Hill, 1978.

[99] Rutherford D and Carter G A, A Heuristic Adaptive Controller for a Sinter Plant, Proc 2nd IFAC Symp Automation in Mining, Mineral and Metal Processing, Johannesburg, 1976.

[100] Shannon C E, Oliver B M and Pierce J R, The Philosophy of Pulse Code Modulation, Proc IRE, Vol 36, pp 1324-1331, 1948.

[101] Shannon C E and Weaver W, The Mathematical Theory of Communication, University of Illinois Press, 1972.

[102] Sharp J A, An Introduction to Distributed and Parallel Processing, Blackwell Scientific Publications, 1987.

[103] Shortliffe E H, Computer Based Medical Consultation: MYCIN, American Elsevier, New York, 1976.

[104] Stallings W, Computer organisation and architectures, Macmillan, New York, USA, 1987.

[105] Sugeno M (Editor), Industrial Applications of Fuzzy Control, North-Holland, 1985.

[106] Sutton R and Towill D R, An introduction to the use of fuzzy sets in the implementation of control algorithms, Jour Inst Elect and Radio Eng, Vol 55, No 10, pp 357-367, 1985.

[107] Timoshenko S, Young D H, and Weaver W, Vibration Problems in Engineering, John Wiley & Sons, New York, 1974.

[108] Thomson W T, Theory of vibrations with applications, 3rd ed, Prentice Hall Englewood Cliffs, NJ, 1988.

[109] Tong R M, A Control Engineering Review of Fuzzy Systems, Automatica, Vol 13, pp 559-569, 1977.

[110] Treu S, Interactive Command Language Design Based on Required Mental Work, Int J Man-Machine Studies, Vol 7, (1), pp 135-149, 1975.

[111] Tse F S, Morse I E and Hinkle R T, Mechanical vibrations, Allyn and Bacon Inc, Boston, 1978.

[112] Van de Vegte J, Feedback Control Systems, Prentice-Hall, New Jersey, 1986.

[113] Van der Rhee F, Van Nauta Lemke and Dijkman J G, Knowledge based fuzzy control of systems, IEEE Trans on Aut Control, Vol 35, No 2, 1990.

[114] Virk G S, Algorithms for State Constrained Control Problems with Delay, Ph D Thesis, Imperial College, University of London, 1982.

[115] Wadsworth C, Who Should Think Parallel ? SERC/DTI Transputer Initiative, Mailshot, Sept 1989.

[116] Warwick K, Control Systems: An Introduction, Prentice-Hall, 1989.

[117] Wiberg D M, State-space and Linear Systems, Schaum's Outline Series, McGraw-Hill, 1971.

[118] Young R M, The Machine Inside the Machine: Users' Models of Pocket Calculators, Int J Man-Machine Studies, 15, pp 15-83, 1981.

[119] Young S J, An Introduction to ADA, Ellis Horwood Publishers, 1983.

[120] Zadeh L A, Fuzzy Sets, Information and Control, Vol 8, pp 338-353, 1965.

[121] Zadeh L A, Outline of a new approach to the analysis of complex systems and decision processes, IEEE Trans on Systems, Man and Cybernetics, SMC-3, No 1, pp 28-44, 1973.

[122] Zadeh L A, Making Computers Think Like People, IEEE Spectrum, pp 26-32, 1984.

[123] Ziegler J G and Nichols N B, Optimum Settings for Automatic Controllers, Trans ASME, Vol 64, pp 759-768, 1942.

Appendix A: Table of Laplace and z-Transforms

Time Function $f(t), \quad t > 0$	Laplace Transform $F(s)$	$z-$Transform $F(z)$
$\delta(t)$	1	1
$\delta(t-kT)$	e^{-ksT}	z^{-k}
Unit step $u_s(t)$	$\frac{1}{s}$	$\frac{z}{z-1}$
Unit ramp, t	$\frac{1}{s^2}$	$\frac{Tz}{(z-1)^2}$
t^2	$\frac{2}{s^3}$	$\frac{T^2 z(z+1)}{(z-1)^3}$
t^{k-1}	$\frac{(k-1)!}{s^k}$	$\lim_{a \to 0} (-1)^{k-1} \frac{\partial^{k-1}}{\partial a^{k-1}} \left[\frac{z}{z-e^{-aT}} \right]$
e^{-at}	$\frac{1}{s+a}$	$\frac{z}{z-e^{-aT}}$
$1-e^{-at}$	$\frac{a}{s(s+a)}$	$\frac{z(1-e^{-aT})}{(z-1)(z-e^{-aT})}$
te^{-at}	$\frac{1}{(s+a)^2}$	$\frac{Tze^{-aT}}{(z-e^{-aT})^2}$

Time Function $f(t),\ t>0$	Laplace Transform $F(s)$	$z-$Transform $F(z)$
$t^k e^{-at}$	$\frac{(k-1)!}{(s+a)^k}$	$(-1)^k \frac{\partial^k}{\partial a^k}\left[\frac{z}{z-e^{-aT}}\right]$
$\frac{1}{(b-a)}\left(e^{-at}-e^{-bt}\right)$	$\frac{1}{(s+a)(s+b)}$	$\frac{1}{(b-a)}\left[\frac{z}{z-e^{-aT}}-\frac{z}{z-e^{-bT}}\right]$
$be^{-bt}-ae^{-at}$	$\frac{(b-a)s}{(s+a)(s+b)}$	$\frac{z\left[z(b-a)-\left(be^{-aT}-ae^{-bT}\right)\right]}{(z-e^{-aT})(z-e^{-bT})}$
$t-\frac{1}{a}(1-e^{-at})$	$\frac{a}{s^2(s+a)}$	$\frac{Tz}{(z-1)^2}-\frac{\left(1-e^{-aT}\right)z}{a(z-1)(z-e^{-aT})}$
$(1-at)\,e^{-at}$	$\frac{s}{(s+a)^2}$	$\frac{z\left[z-e^{-aT}(1+aT)\right]}{(z-e^{-aT})^2}$
$1-e^{-at}(1+at)$	$\frac{a^2}{s(s+a)^2}$	$\frac{z[z\alpha+\beta]}{(z-1)(z-e^{-aT})^2}$ $\alpha=1-e^{-aT}-aTe^{-aT}$ $\beta=e^{-2aT}-e^{-aT}+aTe^{-aT}$
te^{-at}	$\frac{1}{(s+a)^2}$	$\frac{Tze^{-aT}}{(z-e^{-aT})^2}$
$\sin\omega t$	$\frac{\omega}{s^2+\omega^2}$	$\frac{z\sin\omega T}{z^2-2z\cos\omega T+1}$
$\cos\omega t$	$\frac{s}{s^2+\omega^2}$	$\frac{z(z-\cos\omega T)}{z^2-2z\cos\omega T+1}$
$e^{-at}\sin\omega t$	$\frac{\omega}{(s+a)^2+\omega^2}$	$\frac{ze^{-aT}\sin\omega T}{z^2-2ze^{-aT}\cos\omega T+e^{-2aT}}$
$e^{-at}\cos\omega t$	$\frac{s+a}{(s+a)^2+\omega^2}$	$\frac{z^2-ze^{-aT}\cos\omega T}{z^2-2ze^{-aT}\cos\omega T+e^{-2aT}}$

Appendix B: Continuous Second-order Systems

B.1 Introduction

In this appendix we present results relating to continuous second-order systems since they are used widely in conventional control systems design where dominant second-order behaviour can be assumed for higher-order systems. We will assume a transfer function of the following standard form

$$G(s) = \frac{Y(s)}{U(s)} = \frac{\omega_n^2}{s^2 + 2\zeta\omega_n s + \omega_n^2} \tag{B.1}$$

where ζ is defined to be the dimensionless damping ratio, and ω_n is the undamped natural frequency (see Phillips and Harbor [94]; D'Azzo and Houpis [25]; Ogata [88]). Note that the DC gain of the system is unity. The response of this system to a unit step input, subject to zero initial conditions can be shown (see D'Azzo and Houpis [25]) to be given by

$$y(t) = 1 - \frac{e^{-\zeta\omega_n t}}{\sqrt{1-\zeta^2}} \sin\left(\omega_n\sqrt{1-\zeta^2}t + \cos^{-1}\zeta\right) \tag{B.2}$$

A family of curves representing the step responses is shown in Figure B.1, where the horizontal axis is the dimensionless variable $\omega_n t$. The curves are thus functions only of the damping ratio and show that the overshoot is dependent upon ζ; for overdamped and critically damped cases $\zeta \geq 1$, there is no overshoot and no oscillation; for underdamped cases, $0 \leq \zeta < 1$, the system oscillates around the final steady-state value. The peak overshoot as ζ varies is shown more clearly in Figure B.2.

B.2 Response Specifications

Before a control system is designed, specifications must be developed that describe the characteristics that the system should possess. Some of these

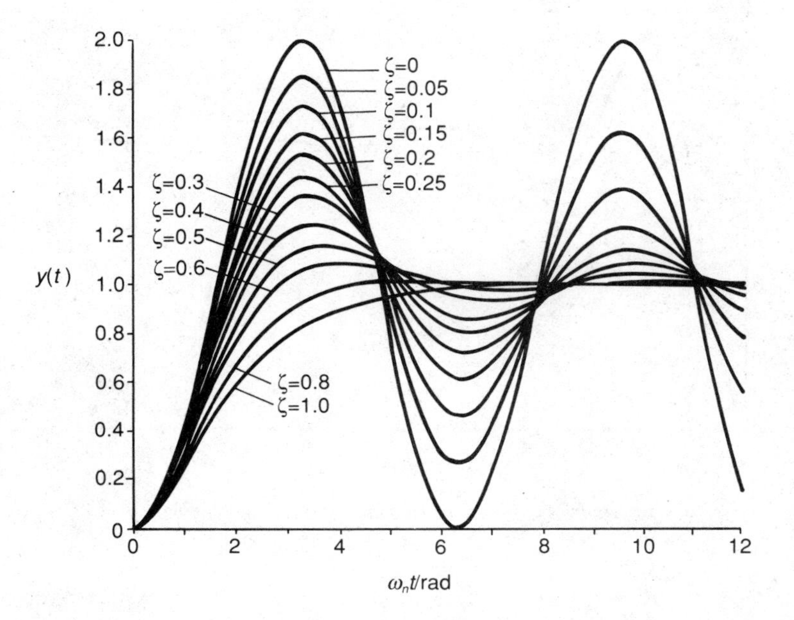

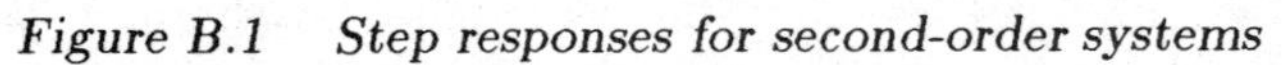

Figure B.1 Step responses for second-order systems

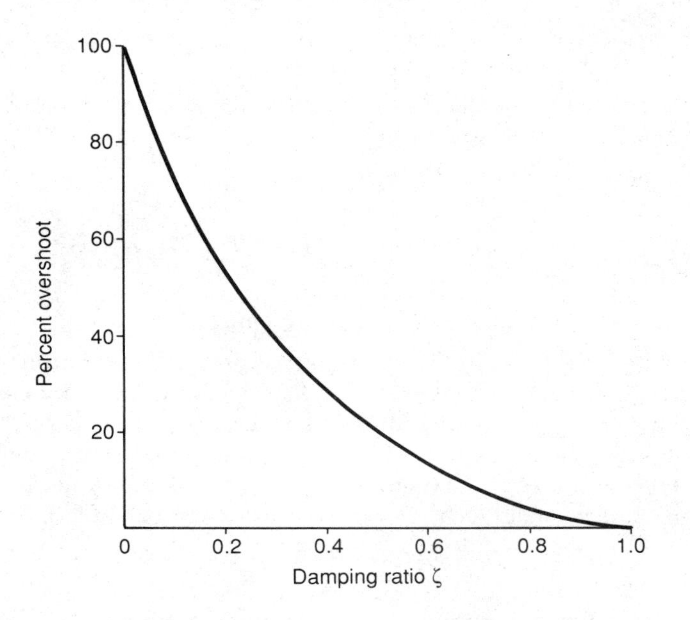

Figure B.2 Peak overshoot versus damping ratio

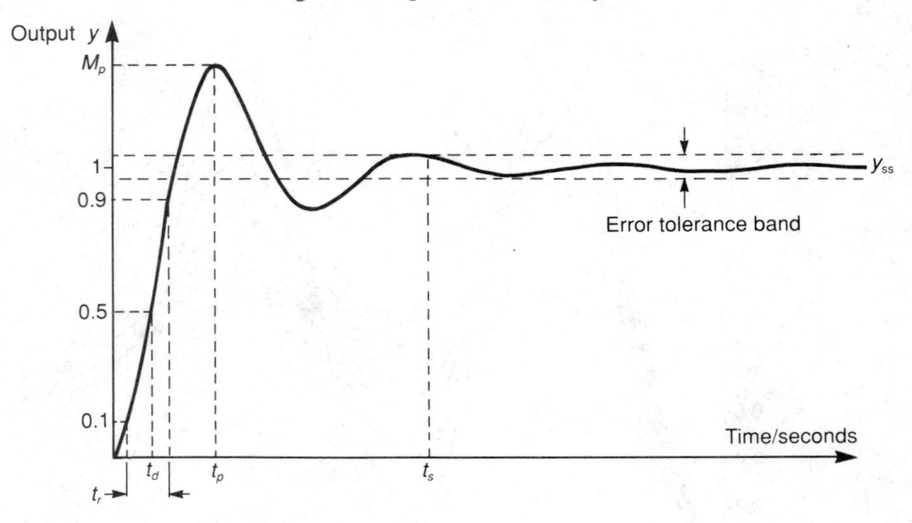

Figure B.3 Unit step response of a system

can be written in terms of the system's step response. A typical step response of a second-order system is shown in Figure B.3. Some characteristics that can be used to describe the response include the following.

(i) The rise time, t_r, is the time taken for the response to rise from 10% of the final value to 90% of the final value.

(ii) The peak value of the step response is denoted by M_p, and the time to reach this peak value is t_p.

(iii) The percentage overshoot is defined by

$$\text{Percentage overshoot} = \frac{M_p - y_{ss}}{y_{ss}} \times 100\% \tag{B.3}$$

where y_{ss} is the final or steady-state value of the output $y(t)$.

(iv) The settling time, t_s, is the time required for the output to settle within a certain percentage of its final value. Commonly used values for the error-band are 2% or 5%. For second-order systems, the value of the transient component at any time is equal to or less than the exponential $e^{-\zeta\omega_n t}$. The settling times expressed in a number of time constants τ for different error bands are given in Table B.1.

(v) The delay time, t_d, is the time required for the response to reach half the final value the very first time.

We have defined the above parameters for the underdamped case; t_r, t_s, y_{ss} and t_d are equally meaningful for the overdamped case, but M_p, t_p and percentage overshoot obviously have no clear meaning in these cases.

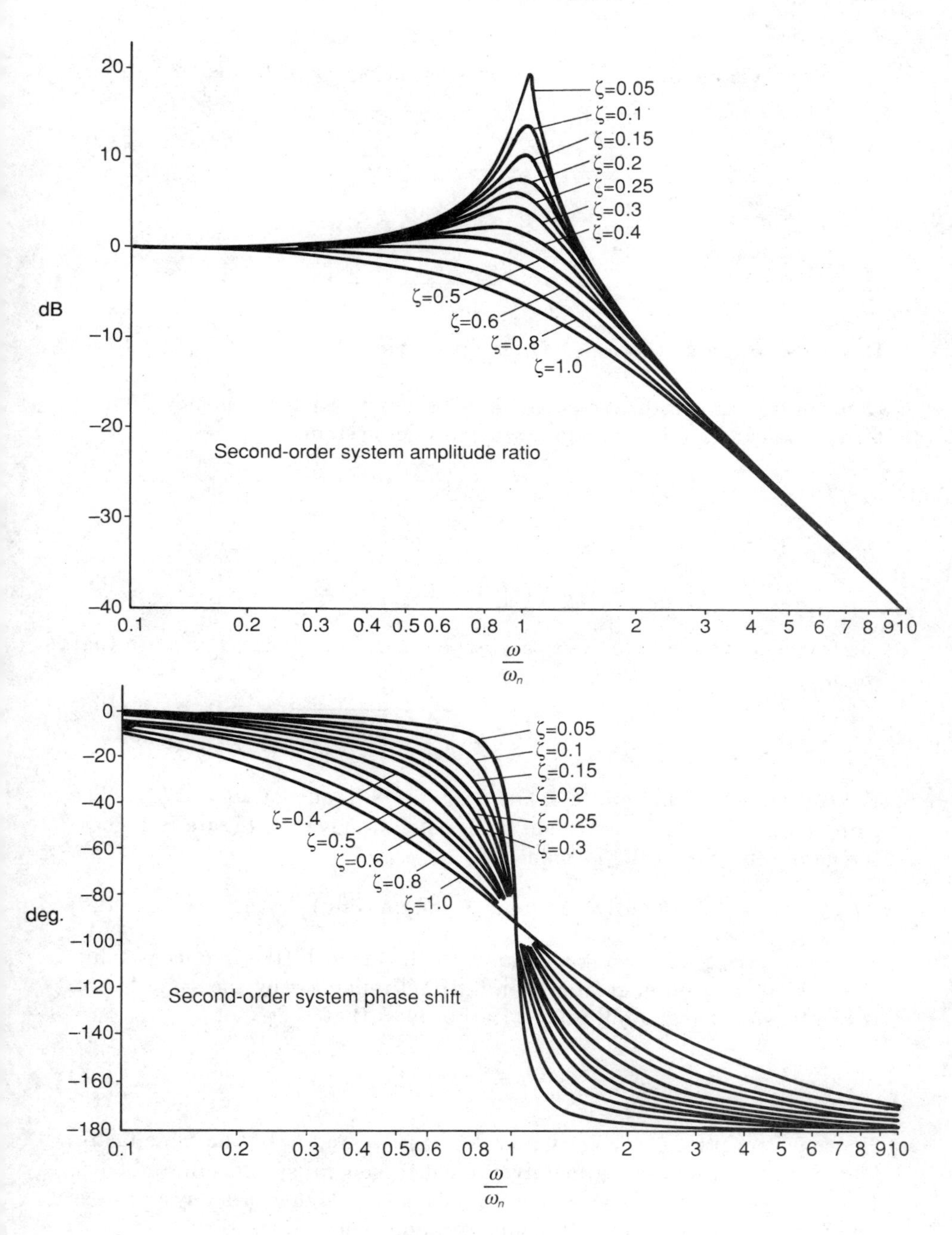

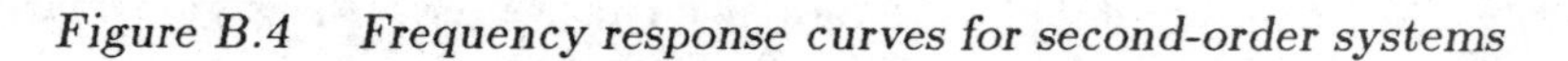

Figure B.4 Frequency response curves for second-order systems

Table B.1 Settling times for various error bands

Error band, %	Settling time t_s, seconds
10	2.3τ
5	3.0τ
2	3.9τ
1	4.6τ

B.3 Frequency Response Characteristics

Control system specifications can also be described in frequency domain terms. Considering the standard second-order system:

$$G(s) = \frac{\omega_n^2}{s^2 + 2\zeta\omega_n s + \omega_n^2} = \frac{1}{\left(\frac{s}{\omega_n}\right)^2 + 2\zeta\left(\frac{s}{\omega_n}\right) + 1} \quad \text{(B.4)}$$

the frequency domain analysis can be performed by letting $s = j\omega$ so that we have

$$G(j\omega) = \frac{1}{\left[1 - \left(\frac{\omega}{\omega_n}\right)^2\right] + j2\zeta\left(\frac{\omega}{\omega_n}\right)} \quad \text{(B.5)}$$

For this transfer function we define normalised frequency $\omega_1 = \omega/\omega_n$. The gain and phase curves for various values of ζ are given in Figure B.4 where the gain in decibels (dB) is defined as

$$\text{dB} = 20 \ \log_{10}(\text{Numeric Gain}) \quad \text{(B.6)}$$

For a constant ζ, increasing ω_n causes the bandwidth (the frequency range over which the gain is greater than 0 dB), to increase by the same factor. It can be shown (see Phillips and Harbor [94]) that

$$\omega_n t_p = \frac{\pi}{\sqrt{1-\zeta^2}} \quad \text{(B.7)}$$

Hence for constant ζ, increasing ω_n decreases t_p and t_r by the same factor. This result can be approximately applied to general systems and not just to second-order ones. The reason for this is that higher-order systems can be approximated by their dominant (second-order) modes.

In the frequency domain we design for concepts such as gain margins and phase margins. These are defined as follows:

Gain Margin

If the magnitude of the open-loop function of a stable closed-loop system at the 180° phase crossover on the frequency response diagram (Nyquist, Bode, etc.) is the value α, the gain margin is $1/\alpha$, and is usually expressed in decibels.

Phase Margin

The phase margin is the magnitude of the angle (180°− phase angle) at the point when the open-loop system gain is unity (0 dB).

The phase margin can be obtained for the standard second-order system as a function of damping ratio ζ and plotted as shown in Figure B.5.

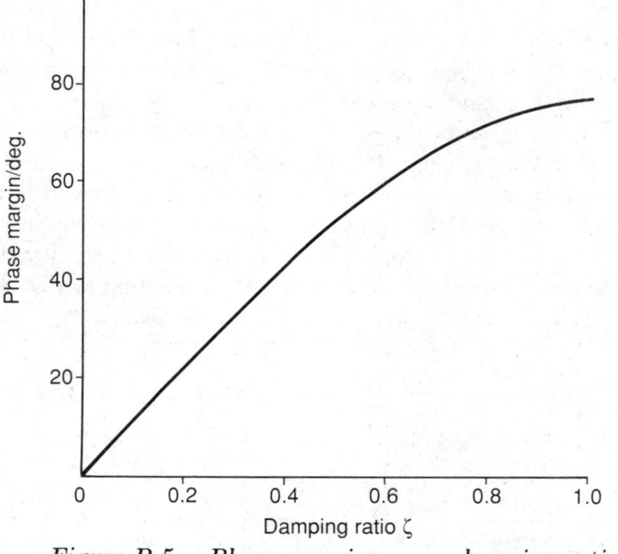

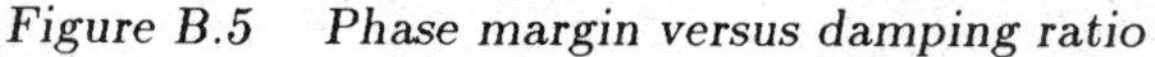

Figure B.5 Phase margin versus damping ratio

As can be seen, the lower end of the graph can be approximated reasonably by the linear relationship

$$\text{Phase Margin} \approx \frac{\zeta}{0.01} \tag{B.8}$$

Appendix C: The Transputer and occam

C.1 Introduction

The general concepts of parallel processing have been discussed in chapter 6 and it is the intention here to concentrate on how such systems can be implemented using transputer technology. The transputer is a VLSI device which comprises a processor, memory and external communication links on a single substrate of silicon. It has been designed by Inmos Ltd to be used as a programmable component in implementing parallel processing systems. In view of this, the word "transputer" has been derived from TRANSistor (a single component that can be used in conjunction with others in electronic circuits), and comPUTER (a machine whose main task

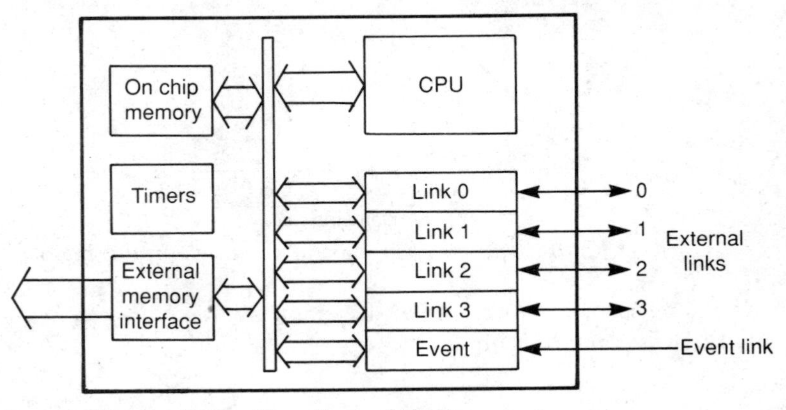

Figure C.1 Structure of the transputer

is processing information/data quickly and efficiently). The basic structure of the transputer is shown in Figure C.1, although there are variations on this to form a family of devices. Transputers come in 16 and 32 bit word-length versions; some have a hardware floating point unit, while others can have disk storage device controllers on chip. The amount of memory and number of links can vary as well as the processing speeds (see Inmos [51] - [56]).

Hence, although a transputer is essentially a computer on a single chip, and can be used as such (with suitable I/O), its unique external links allow other transputers to be connected to it. In this way transputer array systems can be constructed and used in a variety of applications, see IEE Computing and Control Colloquiums [47], [48]; IEE Electronics Colloquium [49]; Freeman and Phillips [38]; Pritchard and Scott [96]. The links allow serial point-to-point communication to take place between processors and thus avoid the bus contention problems in conventional shared memory data-bus computer systems, as discussed in chapter 6. A link can operate at 5, 10, or 20 Mbits per second in both directions at the same time. These speeds are likely to increase with future generations of transputers. Once a link communication has been initiated by the processor, it proceeds autonomously allowing the processor to execute another process.

External interrupt requests can be received on a separate link, called the event link, which is handled in a similar fashion to the communication links. At present there is no support for the prioritisation of interrupt requests and all such matters must be handled explicitly by the system designer. Future transputers are likely to have more event pins, thus allowing for multi-level interrupt structures.

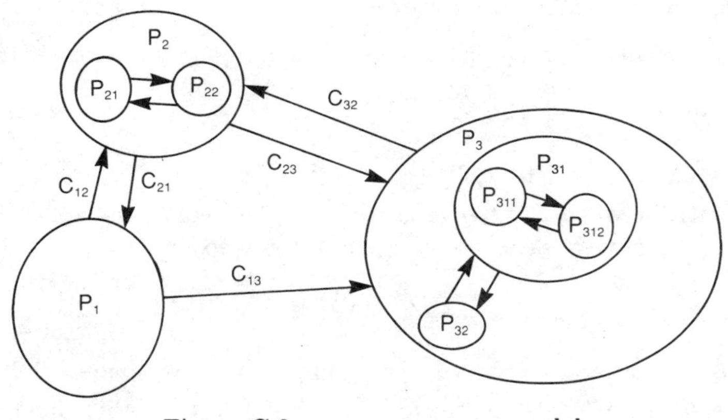

Figure C.2 occam process model

C.2 occam Overview

The model of concurrency supported in hardware by transputers is the occam model, which has the following features:

(i) The world is made up of processes which exist in parallel, that is, at the same time.

(ii) Processes may be born and may die.

(iii) Processes may spawn other processes.

(iv) Processes can communicate between each other using messages through channels.

(v) A channel is a uni-directional, unbuffered link between just two processes.

(vi) A channel provides synchronised communication.

(vii) A process is either a primitive process or a collection of processes.

(viii) A collection specifies both its extent (what processes are in the collection), and how the collection behaves.

A typical occam process model is shown in Figure C.2, where such parallel processes are clearly shown, and how they can interact with each other via one way channels.

Process P_1 is a primitive process which interacts with processes P_2 and P_3, by sending messages to P_2 (via channel C_{12}) and P_3 (via channel C_{13}), while receiving messages only from P_2 (via channel C_{21}). Process P_2 is a collection of communicating processes P_{21} and P_{22}, etc.

Some precise details of occam are now given to illustrate the language; full details can be found in the occam 2 reference manual, [52]. All processes in occam are constructed using three primitive processes, namely:

(i) assignment, when an expression is computed and the result assigned to a variable as in

$$x := 20$$

so that "x" is set to 20;

(ii) input, for when a message is received on a communication channel; the "?" mark is used to indicate input and so

$$\text{In} \; ? \; x$$

sets the variable "x" to the value input from the channel "In";

(iii) output, for when a message needs to be sent to another process. The "!" mark is used to indicate output on a specified channel, and so

$$\text{Out} \; ! \; x$$

ouputs the value "x" on the channel "Out".

Using these three instructions together with the following constructors, more complex processes can be formed. occam uses indentations from the left-hand edge of the page to define the structure of the processes, as will become clear from the following discussion.

SEQ: The sequence constructor defines a process whose component processes are executed in order, terminating when the last process ends, for example:

```
SEQ
  In ? x
  y := x * x
  Out ! y
```

PAR: The parallel constructor defines a process whose component processes are executed concurrently. This process terminates when all of the constituent process have ended, for example:

```
PAR
  Out1 ! a
  Out2 ! b
```

ALT: The alternative constructor defines a process encompassing other processes which have an input as their first component. The first process to become ready is executed. The ALT process terminates when the chosen process terminates, for example:

```
ALT
  In1 ? x
    Out1 ! x
  In2 ? x
    Out2 ! x
```

IF: The condition constructor defines a process, each component of which has a condition as its first component. If a condition is found to be satisfied (that is, TRUE) then that process is executed. The condition process terminates when that process terminates, for example:

```
IF
  x <> 0
    x:= x + 10
  x = 0
    SKIP
```

WHILE: The repetition constructor defines a condition and a process which will be repeatedly executed until a false condition is evaluated, for example:

```
WHILE x <> 0
  SEQ
    In ? x
    Out ! x
```

In addition, all of the above constructors can be repeated for a known number of times, for example:

```
SEQ i = 0 FOR 6
```

```
SEQ
  In ? x
  Out ! x
```

will input "x" on channel "In", and output "x" on channel "Out" 6 times.

The channels are defined in the same way as variables, for example:

CHAN OF INT In: defines the channel "In" to accept variable integer values.

CHAN OF ANY In: allows the channel "In" to accept any type of variable integer, real, or byte.

To enable abstraction, a name can be given to the text of a process, for example:

```
PROC SQU(CHAN OF REAL In, Out)
  REAL16 x:
  SEQ
    In ? x
    Out ! x * x
:
```

Using these constructs, concurrent programs can be written in a straightforward manner. The occam channels provide synchronised communication between two concurrent processes. Hence data is transferred only when the two processes are ready.

Inmos have provided an occam programming environment under the "Transputer Development System" (TDS) where a folding editor is used. Three dots at the start of a line indicate that information is folded away (hidden from view). In this way a hierarchy of folds can be created so that the program has structure and the different levels can be investigated by entering or exiting from folds.

The TDS environment also provides various utilities such as a compiler and configurer. A full description of the facilities can be found in the TDS user guide [53].

C.3 Transputer Systems and occam Configuration

As already mentioned, the transputer has been designed to implement the occam model of concurrency efficiently. Figure C.3 shows a typical transputer system model, showing three transputers communicating through the external links. As shown, this system can be used to implement the occam process model of Figure C.2 in a straightforward way by mapping process P_1 to transputer 1, etc.

A transputer has its own scheduler and so it can be used to run a number of concurrent processes together by time-sharing the processor. Also, the

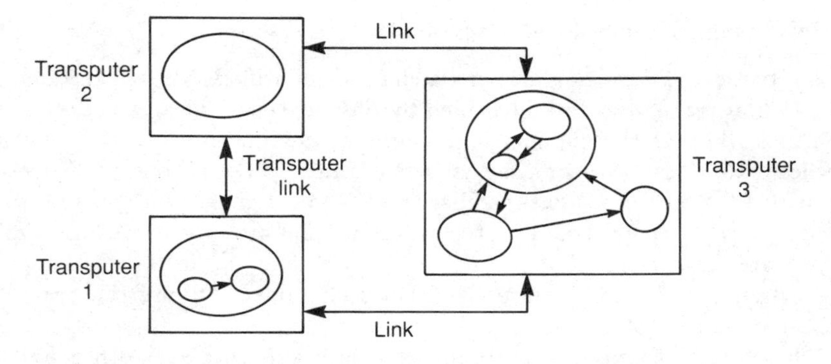

Figure C.3 Transputer system model

communication channels between processes can be through the external hardware links on the transputers or internal software channels by using memory locations. Each transputer link provides two occam channels; one channel is an input channel, and the other an output channel. Both have to be specified in the program.

The external link channels between transputers and the internal software channels on one transputer are seen to be identical at run-time. Therefore it is possible to develop a solution to a problem independently of the actual transputer network upon which it will be executed. For instance the solution can be developed on a single transputer, and once the solution is functional, the various processes can be allocated to different transputers, and the channels allocated to the links of the appropriate transputers for final implementation. The occam configuration facilities that allows such allocations are as follows:

PLACED PAR:

This is similar to the ordinary *PAR* statement except that the processes are to be placed on separate transputers.

PROCESSOR number transputer.type:

This is used to identify a *number* to a processor and what *transputer.type* the processor is, so that the configurer can check that the correct compiler was used, that is, T2 or T212 for the T212, T222, T225 or M212 16-bit transputers, T4 or T414 for the T414 32-bit transputer, T425 for the T425 32-bit transputer and T8 or T800 for the T800, T801 or T805 32-bit floating point transputers.

PLACE channel.name AT link.address:

This provides a means of allocating a channel identified as *channel.name* to a particular transputer link identified by *link.address*. The *link.address* also identifies whether the channel is a source or destination by the following method: the four links on, say, a T414 transputer give rise to 8 occam channels (4 input and 4 output channels). Link 0 gives channels 0 and 4, Link 1 gives channels 1 and 5 etc., where 0,1,2,3 are output channels and 4,5,6 7 are input channels.

Further details are given in the TDS user guide, [53] and the occam 2 reference manual [52].

Therefore transputers and occam are ideally suited to each other, and in fact Inmos originally intended all transputer programming to be performed in occam. A claim made by Inmos is that the occam compiler produces almost optimal executable code which can only be marginally improved by an expert assembly language programmer. An average assembly language programmer would produce worse code than the compiler. In spite of this claim, which appears to be supported by many computer scientists, there has been a strong resistance to occam from the application user community because it is a new language which has to be learnt and implemented. Many millions of man years have been spent on writing programs in a variety of languages – for these to be run on transputer systems would require major re-writing and expenditure. Therefore, the majority of users required new compilers to support their normal programming languages on transputer hardware so that their existing codes could be executed on transputers. In the light of these demands, several high level language compilers for transputers have been developed, and are now available. Parallel extensions to the sequential languages have also been inserted to give parallel versions of the languages. It is now possible to program multi-transputer systems in Fortran, C, Pascal, etc. Software tools to assist users in programming and debugging parallel computer systems are also emerging.

We now present an example which will illustrate the parallel programming approach using transputers and occam.

C.4 Vibration Control of a Flexible Cantilever

The example we will consider is that of vibration suppression in a flexible cantilever system. Problems of vibration are common in engineering applications and there is a clear need for these problems to be studied so that effective control/vibration suppression techniques can be developed and implemented. Examples of where such oscillatory behaviour occurs include aircraft fuselage and wings, satellite solar panels and buildings. The problems stem mainly from the use of lightweight designs aimed at improving performance and/or reducing costs, but which also lead to vibrational difficulties. The main methods for handling the unwanted vibrations are:

(i) to introduce passive damping elements such as springs and/or dampers,

(ii) to use active control techniques to provide the damping actively when required, or

(iii) a combination of the two techniques.

We will use the active damping method to address and computer-control, in real-time, the vibrational problems encountered in a flexible beam structure. We shall concentrate on a cantilever system which can represent, for example, vibrations in aircraft wings and/or the swaying of tall skyscrapers in windy conditions. Other applications can be considered by changing the boundary conditions on the beam – for example free–free beams can be used to represent aircraft fuselage oscillations and hinged–free beams for satellite solar panel oscillations. Clearly the simple beam system will not include all the two- and three-dimensional vibrational effects, but it does represent a first attempt for the consideration of several practical applications.

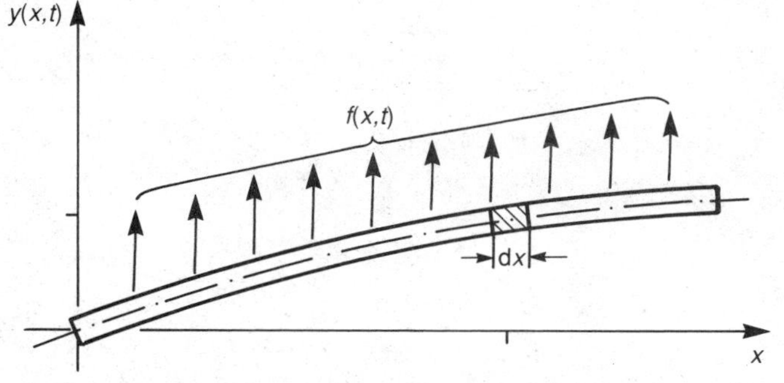

Figure C.4 Cantilever system in transverse vibration

A cantilever system in transverse oscillation (see Figure C.4) has its motion described by the fourth-order partial differential equation (PDE), see for example Timoshenko *et al.* [107],

$$\mu^2 \frac{\partial^4 y(x,t)}{\partial x^4} + \frac{\partial^2 y(x,t)}{\partial t^2} = f(x,t) \tag{C.1}$$

where $y(x,t)$ is the deflection at a distance x from the fixed end at time t, μ is a beam constant and $f(x,t)$ is a force causing the beam to deform. It is well known that flexible structure systems such as those described by equation (C.1) have an infinite number of modes, see Meirovitch [80], Newland [86], Thomson [108] and Tse *et al.* [111], although in most cases the lower order modes are the dominant ones which need consideration. Higher modes have little effect in this beam system and can be ignored in much of the analysis, see Kourmoulis [68]. However what is a low mode and what is a high mode is a subjective decision, although it is clear that as more modes are considered in the analysis, the better is the performance that one can expect. Such an increase in dimensionality leads to requiring vast computing resources so that the real-time processing demands are satisfied, for the modelling and controller design and implementation. In fact the computing demands can get so enormous in practice that they are difficult to satisfy using sequential computing methods. Therefore parallel processing methods using transputer hardware, as considered here, can be more appropriate.

The approach taken here follows along the lines outlined in chapter 6 where the overall computing task is divided into the following main sub-tasks:

(a) System simulation (necessary if the actual system is unavailable, does not exist or for testing of the modelling and controller designs).

(b) Controller design, implementation and assessment.

(c) Modelling (generally reduced order) for state estimation and validation.

(d) User interface.

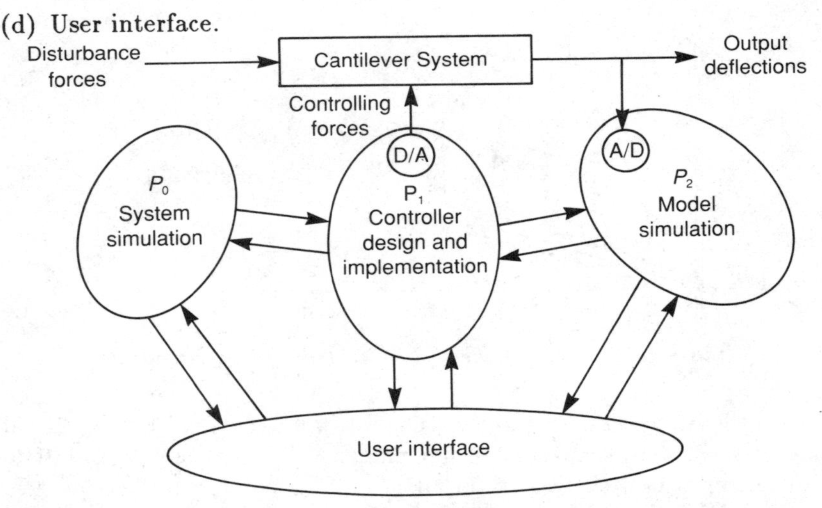

Figure C.5 occam process model for cantilever control

These sub-tasks can be treated as separate occam processes which interact with each other to give the overall solution. One such design can be as shown in Figure C.5. An obvious approach for solving the problem is where the three processes $P_0 - P_2$ are mapped onto three transputers as shown in Figure C.6, together with a master processor which controls the overall computations and performs the user interface through a personal computer.

In programming multi-transputer systems of this kind using the TDS, there are two sections of code that need to be written. The first of these

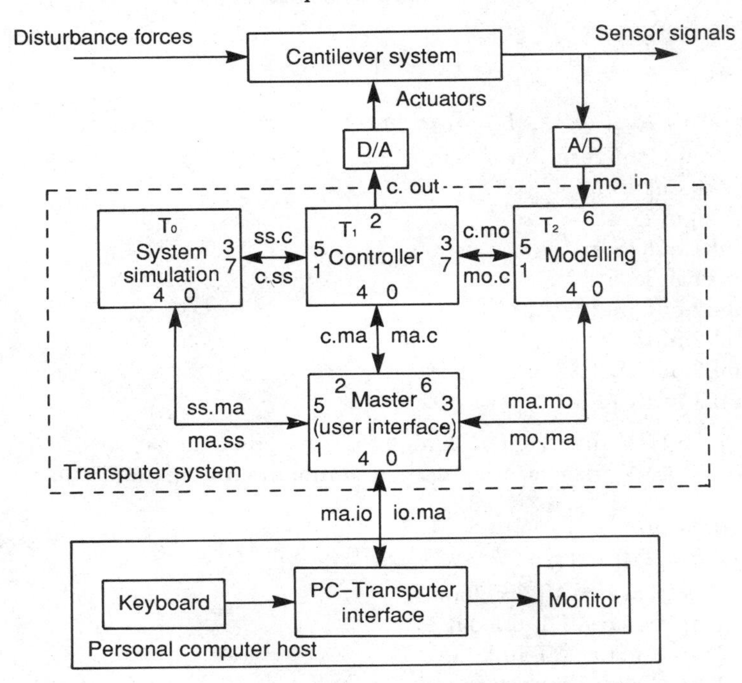

Figure C.6 Transputer system for cantilever control

is the "EXE", which is the code that is executed by the master processor (also known as the root processor), and the second is the "PROGRAM", which is the code that is executed by the transputer network connected to the master transputer. The configuration, that is, how the transputers are interconnected, needs to be specified. The root transputer's connections are defined in the "EXE", and the remaining transputer linkages are specified in the "PROGRAM". For example, the configuration fold of the three transputers $T_0 - T_2$ is shown in Table C.1. This fold would be preceded by the separately compiled codes for the processes $P_0 - P_2$ in the "PROGRAM" fold. The "EXE" and "PROGRAM" folds, partly unfolded, are shown in Table C.2. The "SC" after the fold dots signifies a separately compiled process (see the TDS user guide [53]).

To execute the overall solution, both the transputer network and the root transputer need to be loaded with their codes and configurations. The 'PROGRAM" code is first configured onto the three transputers $T_0 - T_2$. Once this has been done the three transputers have their respective programs and commence execution. The root transputer then has the "EXE" code loaded and run. Since occam synchronises all communications, the overall solution can therefore only be available when all four transputers (including the root) have their programs and are executing them.

Table C.1 Transputer configuration fold

```
{{{ System Configuration
VAL link0.out IS 0:
VAL link0.in IS 4:
VAL link1.out IS 1:
VAL link1.in IS 5:
VAL link2.out IS 2:
VAL link2.in IS 6:
VAL link3.out IS 3:
VAL link3.in IS 7:
CHAN OF ANY mo.in, c.out, ss.c, c.ss, c.mo, mo.c, c.ma:
CHAN OF ANY ma.c, ss.ma, ma.ss, ma.mo, mo.ma, ma.io, io.ma :

PLACED PAR
  PROCESSOR 0 T8
    PLACE ss.ma AT link0.out :
    PLACE ss.c AT link3.out :
    PLACE ma.ss AT link0.in :
    PLACE c.ss AT link3.in :
    system.simulation(ss.ma, ss.c, ma.ss, c.ss)

  PROCESSOR 1 T8
    PLACE c.ma AT link0.out :
    PLACE c.ss AT link1.out :
    PLACE c.out AT link2.out :
    PLACE c.mo AT link3.out :
    PLACE ma.c AT link0.in :
    PLACE ss.c AT link1.in :
    PLACE mo.c AT link3.in :
    controller(c.ma, c.ss, c.out, c.mo, ma.c, ss.c, mo.c)

  PROCESSOR 2 T8
    PLACE mo.ma AT link0.out :
    PLACE mo.c AT link1.out :
    PLACE ma.mo AT link0.in :
    PLACE c.mo AT link1.in :
    PLACE mo.in AT link2.in :
    modelling(mo.ma, mo.c, ma.mo, c.mo, mo.in)
}}}
```

Table C.2 Cantilever control TDS fold structure

```
{{{   Control example

{{{ ... EXE user interface
{{{ F  user interface
...      link declarations
...      SC PROC user.interface
CHAN OF ANY ma.io, ma.ss, ma.c, ma.mo, io.ma, ss.ma, c.ma, mo.ma :
PLACE ma.io AT link0.out :
PLACE ma.ss AT link1.out :
PLACE ma.c AT link2.out :
PLACE ma.mo AT link3.out :
PLACE io.ma AT link0.in :
PLACE ss.ma AT link1.in5 :
PLACE c.ma AT link2.in :
PLACE mo.ma AT link3.in :
user.interface(ma.io, ma.ss, ma.c, ma.mo, io.ma, ss.ma, c.ma, mo.ma)
}}}
}}}

{{{ PROGRAM network
{{{ F network
...      link declarations
...      SC PROC system.simulation
...      SC PROC controller design and implementation
...      SC PROC modelling
...      system configuration
}}}
}}}

}}}
```

In this way a large complex computational problem can be broken down into smaller subproblems which can be solved on a network of transputers in an efficient manner. Kourmoulis [68] presents real-time simulation results for this cantilever control example where the system simulation, modelling, controller design and user interface sub-tasks are solved on tailor-made transputer architectures. The complete network is shown in Figure C.7, and consists of 17 T800 and 2 T414 transputers.

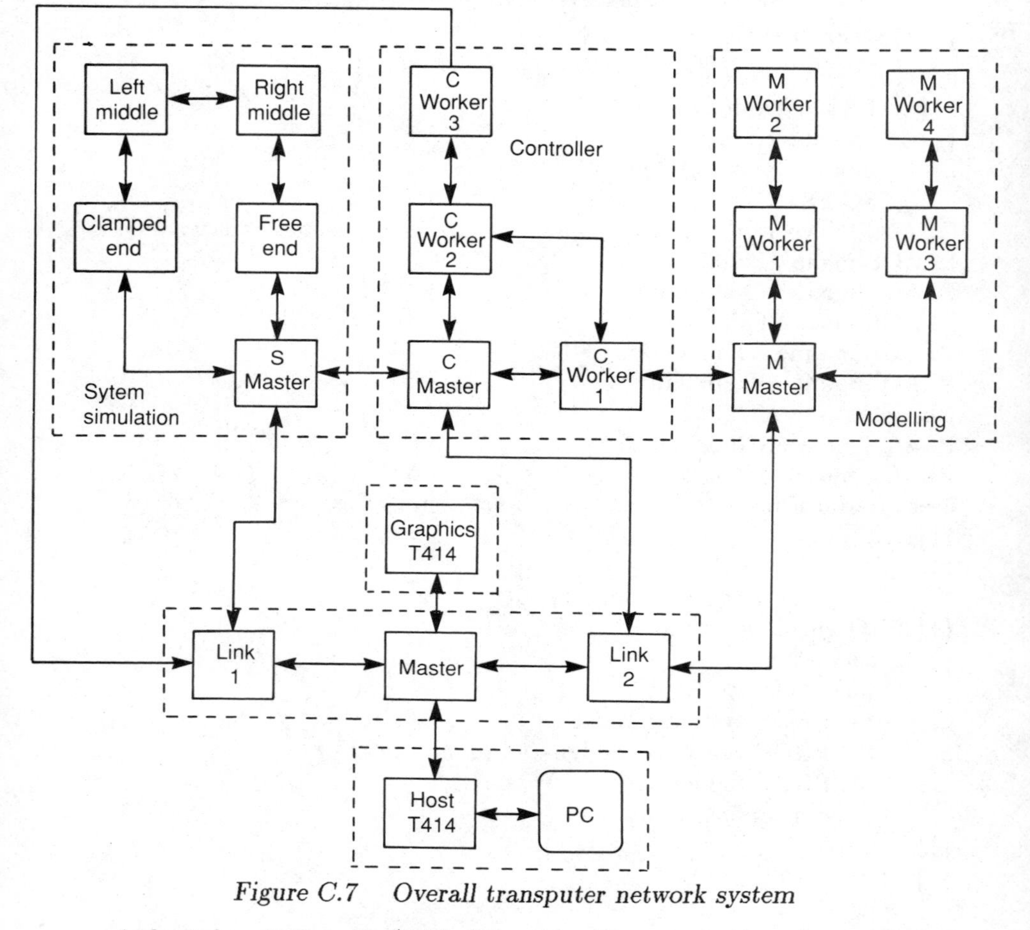

Figure C.7 Overall transputer network system

A short description of each block is necessary to appreciate the approach; a fuller discussion can be found in Kourmoulis [68].

System Simulation

To simulate the cantilever system, the PDE in equation (C.1) can be numerically solved using finite difference approximations in time and space.

The beam (of length 63.5 cm) is partitioned into 20 sections which can be split over four processors and using a time stepsize of 0.3 ms, real-time simulation is possible. Timing comparisons, on various computer hardware, are shown in Table C.3.

Table C.3 System and Modelling simulation timings

System simulation	
Computer hardware	Processing time (s)
SUN 3	519.42
$1 \times T414$	1352.75
$1 \times T800$	131.08
$4 \times T800$	40.46

Model simulation	
Computer hardware	Processing time (s)
SUN 3	44.29
$1 \times T414$	155.82
$1 \times T800$	13.81
$5 \times T800$	4.26

Actual real-time over which simulations are performed = 60 seconds

Modelling and State Estimation

Since the infinite number of modes in equation (C.1) are difficult to handle, a lower-order model is derived for use in controller design and implementation. Assuming that the first five modes are dominant and adequate for our needs, a well structured 10th order state-space representation for the cantilever can be easily obtained, see Inman [50]. Here the individual modes are decoupled from each other which allows straightforward parallel implementation on different processors, while keeping communication overheads low and so allows good speedup. The state-space equations can be distributed onto five transputers and then solved using the same input as that applied to the beam system to yield performances down to 7% of real-time.

The model simulation thus obtained can provide estimates of the dominant states which can therefore be used in control law implementation, such as in state feedback compensation for example. Such a model simulation is essentially an open form of state observer which has initialisation difficulties. A better approach is to use a closed form of observer where the system and observer outputs are compared to provide an additional corrective signal to drive the state estimator as discussed in chapter 4.

Controller Design

The beam system PDE of equation (C.1) has zero damping hence any disturbance will set it into continual oscillation. This is clearly not a desirable situation and damping needs to be inserted using a suitable controller. Many methods can be used for the controller design such as pole-placement, PID, dead-beat and optimal control. The approach taken here is to use state feedback (see Kourmoulis [68]) to introduce critical damping to each of the decoupled modes considered, leaving the natural frequencies unchanged. A controller is designed for the 10th-order model and applied to both the model and the system simulation blocks on-line. Figure C.8 shows a typical response of the model and system simulation blocks under state feedback when a disturbing step force is applied. Table C.4 shows the overall timing performance of the system shown in Figure C.7 as well as timings for other hardware implementations.

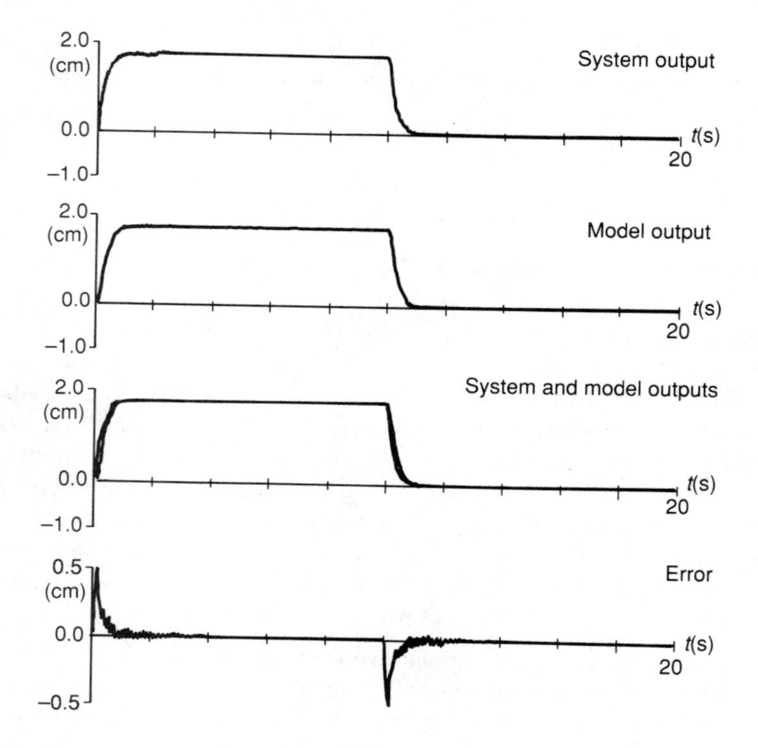

Figure C.8 State feedback control results of cantilever system

Table C.4 Timings of vibration control transputer system

Actual time of simulation (s)	Network Processing Time (s)			
	SUN 3	1 × $T414$	1 × $T800$	17 × $T800$ 2 × $T414$
1.0	26.44	81.96	7.55	0.68
30.0	793.29	2538.53	214.4	18.97
60.0	1586.57	5235.68	453.31	40.82

User Interface

In most control applications, it is useful to have a graphics screen to display the performance of the system. In this respect such a display was developed and included the ability to display the deflection of any point along the beam with time, error displays and real-time animation of the beam.

Appendix D: Solutions to Problems

Chapter 2

1. (i) For $u(t) = 1,\ t \geq 0$

$$\begin{aligned} U(z) &= 1 + z^{-1} + z^{-2} + z^{-3} + \cdots \\ &= \frac{z}{z-1} \end{aligned}$$

(ii) For $u(t) = e^{-at},\ t \geq 0$

$$\begin{aligned} U(z) &= 1 + e^{-aT} z^{-1} + e^{-2aT} z^{-2} + e^{-3aT} z^{-3} + \cdots \\ &= \frac{z}{z - e^{-aT}} \end{aligned}$$

(iii) For $u(t) = \sin \omega t,\ t \geq 0$

$$\begin{aligned} U(z) &= \sin \omega T z^{-1} + \sin 2\omega T z^{-2} + \sin 3\omega T z^{-3} + \cdots \\ &= \frac{z \sin \omega T}{z^2 - 2z \cos \omega T + 1} \end{aligned}$$

(iv) For $u(t) = \cos \omega t,\ t \geq 0$

$$\begin{aligned} U(z) &= 1 + \cos \omega T z^{-1} + \cos 2\omega T z^{-2} + \cos 3\omega T z^{-3} + \cdots \\ &= \frac{z^2 - z \cos \omega T}{z^2 - 2z \cos \omega T + 1} \end{aligned}$$

2. (i) $y(kT) = 1.333 - 1.333(-0.5)^k$, for $k = 0, 1, 2, 3, \ldots$.

(ii) $y(kT) = \begin{array}{cc} -2(0.707j)^k & \text{for } k \text{ odd} \\ 0 & \text{for } k \text{ even} \end{array} \Bigg\}$ for $k = 0, 1, 2, 3, \ldots$.

(iii) $y(kT) = 2 - 2(-1)^k$, for $k = 0, 1, 2, 3, \ldots$.

(iv) $y(kT) = 6.62 - 14.65(0.734)^k + 7.5(0.614)^k$, for $k = 0, 1, 2, 3, \ldots$.

3. The closed-loop transfer function is

$$\frac{Y(z)}{U(z)} = \frac{KT}{z - (1 - KT)}$$

For $K = 1$, and unit step input

$$y(kT) = 1 - (1 - T)^k, \quad \text{for } k = 0, 1, 2, 3, \ldots$$

Hence the output depends on the sampling interval T.

$T = 0.25$ s

$y(kT) = 1 - (0.75)^k$, for $k = 0, 1, 2, 3, \ldots$ which gives an overdamped response. This holds for $T = 0.25 \to 1^-$.

$T = 1$ s

$y(kT) = 1$, for $k = 1, 2, 3, \ldots$ which gives a deadbeat response. As T is further increased the system becomes underdamped.

$T = 1.5$ s

$y(kT) = 1 - (-0.5)^k$, for $k = 0, 1, 2, 3, \ldots$, which gives an oscillatory response. As T is further increased the oscillations become larger as the damping is reduced.

$T = 2$ s

$y(kT) = 1 - (-1)^k$, for $k = 0, 1, 2, 3, \ldots$, which gives continual oscillations. The system is critically stable at this point. As T is further increased the system goes unstable. For example, consider:

$T = 3$ s

$y(kT) = 1 - (-2)^k$, for $k = 0, 1, 2, 3, \ldots$, which gives oscillations that grow in magnitude.

4. The limiting values are $K \geq 9$ and $T \leq 0.26$ s.
5. (i) The z-transform of the output is

$$Y(z) = \frac{0.0049834z + 0.0049668}{z^2 - 1.985z + 0.995} U(z)$$

(ii) The output response is

$$\begin{aligned}
y(0.1) &= 0.00498 & y(1.7) &= 1.097 \\
y(0.2) &= 0.01984 & y(1.8) &= 1.19 \\
y(0.3) &= 0.044 & y(1.9) &= 1.282 \\
y(0.4) &= 0.0783 & y(2.0) &= 1.37 \\
y(0.5) &= 0.1212 & y(2.1) &= 1.454 \\
y(0.6) &= 0.1727 & y(2.2) &= 1.533 \\
y(0.7) &= 0.2321 & y(2.3) &= 1.606 \\
y(0.8) &= 0.2988 & y(2.4) &= 1.672 \\
y(0.9) &= 0.3722 & y(2.5) &= 1.732 \\
y(1.0) &= 0.4514 & y(2.6) &= 1.784 \\
y(1.1) &= 0.5357 & y(2.7) &= 1.828 \\
y(1.2) &= 0.6242 & y(2.8) &= 1.863
\end{aligned}$$

$$\begin{aligned} y(1.3) &= 0.7159 & y(2.9) &= 1.889 \\ y(1.4) &= 0.8099 & y(3.0) &= 1.906 \\ y(1.5) &= 0.9054 & y(3.1) &= 1.914 \\ y(1.6) &= 1.001 & y(3.2) &= 1.913 \end{aligned}$$

The maximum overshoot is $\approx 91\%$, hence the system is very lightly damped.

(iii) The final value theorem gives the steady-state value of $y(kT)$ as 1.

6. The modified z-transform of $G(z)$ is given by

$$G(z,m) = \frac{z^2 a_2 + z a_1 + a_0}{z(z-1)(z-0.000977)}$$

where

$$\begin{aligned} a_0 &= 0.000677m - 0.0005793 + 0.1e^{-6.931m} \\ a_1 &= 0.7932 - 0.694m - 0.2e^{-6.931m} \\ a_2 &= 0.6931m - 0.1 + 0.1e^{-6.931m} \end{aligned}$$

and so the output $Y(z,m)$ for a unit step input is

$$Y(z,m) \frac{z^2 a_2 + z a_1 + a_0}{z^3 - 1.4078z^2 + 0.508z - 0.1}$$

(i) The z-transform of the output is

$$Y(z) = \frac{0.59344z + 0.0992}{z^3 - 1.4078z^2 + 0.508z - 0.1}$$

(ii) For $m = 0.5$, $y(kT, m)$ is as follows

$$\begin{aligned} y(0.5T) &= 0.2497 & y(3.5T) &= 1.013 \\ y(1.5T) &= 0.7915 & y(4.5T) &= 1.004 \\ y(2.5T) &= 0.9874 & y(5.5T) &= 0.9978 \end{aligned}$$

(iii) The output response $y(kT)$, from $Y(s)$, equals

$$\begin{aligned} y(0) &= 0 & y(4T) &= 1.012 \\ y(T) &= 0.5932 & y(5T) &= 1.004 \\ y(2T) &= 0.9344 & y(6T) &= 1.0 \\ y(3T) &= 1.014 & y(7T) &= 0.9998 \end{aligned}$$

Chapter 3

1. The system is unstable, and the z-domain poles are at $-0.215 \pm j1.41$.

2. (i) Two unstable poles, one stable pole and none on the boundary. The z-domain poles are at 4.164, 1.228 and -0.391.
   (ii) Two unstable poles, one stable pole and none on the boundary. The z-domain poles are at $1.144 \pm j1.580$ and -0.789.
   (iii) No unstable poles, two stable poles and two poles on the boundary. The z-domain poles are at $0.5 \pm j0.866, 0.5$ and -0.6.
   (iv) Two unstable poles, two stable poles and none on the boundary. The z-domain poles are at $0.158 \pm j0.063$ and $-1.092 \pm j0.434$.
   (v) One unstable pole, no stable poles and two poles on the boundary. The z-domain poles are at $-0.6 \pm j0.8$ and -2.4.
   (vi) Two unstable poles, two stable poles and two poles on the boundary. The z-domain poles are at $-2, -1.5, \pm j, -0.6$ and 0.5.
3. The closed-loop transfer function is

$$\frac{Y(z)}{U(z)} = \frac{G_1(z)\,G_2(z)}{1 + G_1(z)\,(G_2H)(z)}$$

   When a unit step input is applied

$$\begin{aligned} y(kT) &= 0.295\delta(t-T) + 0.811\delta(t-2T) + 1.338\delta(t-3T) \\ &\quad +1.682\delta(t-4T) + 1.74\delta(t-5T) + 1.524\delta(t-6T) \\ &\quad +1.142\delta(t-7T) + 0.752\delta(t-8T) + \cdots \end{aligned}$$

   which has a steady state value of unity.
4. $K \le 11.09$ for stability. For $K = 10$ the system has a phase margin of $\approx 19.5°$.
5. The z-transform of the output of ZOH_2 is

$$\theta_o(z) = \frac{0.19z^3 + 0.15z^2}{z^4 - 1.93z^3 + 1.69z^2 - 0.9z + 0.15}$$

   Using long division

$$\theta_o(z) = 0.19z^{-1} + 0.52z^{-2} + 0.68z^{-3} + \cdots$$

   and so the output equals 0 at $t = 0$, 0.19 at $t = T$, 0.52 at $t = 2T$, etc. For $t = \infty$, we use the final value theorem to give

$$\lim_{t\to\infty} \theta_o^*(t) = \lim_{z\to 1} \frac{z}{z-1}\theta_o(z) = 0.5$$

6. $K \approx 2.02$ when damping ratio is zero. When $K = 1$, the damping ratio ≈ 0.23 and the percentage overshoot $\approx 57.3\%$.
7. $K = 1$ is the limiting value for stability.
8. $0 \le K \le 378.75$ for stability.

Chapter 4

1. (i) System is uncontrollable, unobservable and unstable (poles are at $z = 0, -2$ and 0.5).
   (ii) System can be stabilised because the unstable mode is controllable. With output feedback, we need $1 \leq k \leq 3$ for stability.
2. Designs can proceed along many lines.
   For example we can use the root-locus method in the s-plane to get $D(s)$ and digitise this to get $D(z)$. It is obvious that for zero steady-state error to a step input, the compensated system must be at least type 1, we therefore need to introduce an integrator into the open-loop transfer function. To keep system second-order, cancel the pole at $s = -1$. Therefore proposed controller is

$$D(s) = K\frac{(s+1)}{s}$$

   This gives a root-locus with second-order asymptotes along $-7.5 \pm j\alpha$. To satisfy the damping ratio requirements $\alpha \approx 13$ which corresponds to an undamped natural frequency $\omega_n \approx 15$ rad/s. The settling time requirements (for 2.5% error band) are given by $t_s = 4/\zeta\omega_n \rightarrow \omega_n = 8$ rad/s. Hence above ω_n of 15 rad/s will be adequate. From the closed-loop compensated transfer function, we can deduce that $K = 225$ gives the required damping ratio of 0.5. However since $D(s)$ will be digitised, and this will introduce errors, we choose $K = 100$ to allow a safety margin. Therefore

$$D(s) = 100\frac{(s+1)}{s}$$

   This needs to be digitised after selecting a suitable sampling interval as discussed in the text. Then controlled performance can be implemented, assessed, etc.
3. The optimal solution is

$$u^*(1) = -9.8, \qquad x^*(2) = \begin{bmatrix} 30.2 \\ -34.6 \end{bmatrix}$$

$$u^*(2) = 56.8, \qquad x^*(3) = \begin{bmatrix} 17.9 \\ -11.6 \end{bmatrix}$$

   If there is no weighting on the control term, the optimisation procedure will drive x_1 and x_2 to zero as quickly as possible without any regard to how much control action is used.
4. The system is unstable without compensation. We will design $D(s)$ and digitise to get $D(z)$. The specifications are given in the time domain, so it is convenient to use the root-locus method to design $D(s)$.

Drawing the uncompensated root-locus (see Figure D.1(a)) shows (not to scale) that we need to insert a zero near the double pole at the origin to pull the locus over to the left half plane. A P+D controller will do but this will not be realisable in digital terms, and so we will use a lead-lag controller whose pole is far to the left so that it has negligible effect upon the dominant portion of the root-locus.
The $\zeta = 0.5$ specification means that the dominant closed-loop poles need to lie on this ζ line. Hence the poles are required to be at $\alpha(-1 \pm j1.73)$ where α is some scalar. The settling time $t_s \leq 1$ s gives, from appendix B, that $\omega_n = 8$ for a 2.5% error band. The closer the compensator zero is to the double pole, the more significant its effect. We will try putting the zero at -1 and the pole at -40 (far into the left-half plane), and so the proposed controller is

$$D(s) = K\frac{(s+1)}{(s+40)}$$

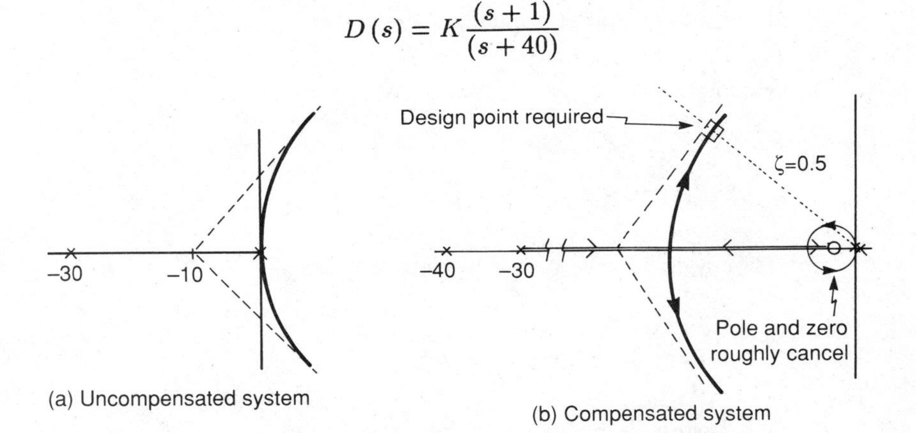

Figure D.1 Magnetic suspension compensation design

This gives the root-locus shown in Figure D.1(b) (not to scale), and a characteristic equation of

$$s^4 + 70s^3 + 1200s^2 + 30Ks + 30K = 0$$

We need to determine the value of K that gives the desired design point on the locus (a CAD package is useful for this). K is in fact found to be ≈ 13. The resulting $D(s)$ needs to be digitised; we will use the pole-zero mapping, with a quite short sampling interval $T = 0.01$ s for good accuracy.
$D(z)$ has a zero at $e^{-T} = 0.99$, and a pole at $e^{-40T} = 0.67$. Matching DC gains gives

$$13\frac{(s+1)}{(s+40)}\Big|_{s=0} = \bar{K}\frac{(z-0.99)}{(z-0.67)}\Big|_{z=1}$$

Hence

$$D(z) = 10.7\frac{(z-0.99)}{(z-0.67)}$$

This can be implemented as follows:

$$D(z) = \frac{m_t}{e_t} = 10.7\frac{(z-0.99)}{(z-0.67)}$$

where m_t represents the signal applied to the system at the sampling instant time t, and e_t is the error measured by the computer at time t. Dividing the top and bottom of the right-hand side by z, cross multiplying and rearranging gives

$$m_t = 0.67m_{t-T} + 10.7(e_t - 0.99e_{t-T})$$

as the signal that is applied to the system.

5. $F = [1.7 \quad 3.2 \quad -0.4]$.
6. Continuous state-space equation is

$$\begin{aligned} \dot{x}(t) &= \begin{bmatrix} -1 & 0 \\ 0 & -2 \end{bmatrix} x(t) + \begin{bmatrix} -1 \\ 2 \end{bmatrix} u(t) \\ y(t) &= \begin{bmatrix} 1 & 1 \end{bmatrix} x(t) \end{aligned}$$

When a unit input step is applied we have

$$\begin{aligned} y(T) &= 0.25 \\ y(2T) &= 0.188 \\ y(3T) &= 0.109 \\ y(\infty) &= 0. \end{aligned}$$

7. (i) System is controllable, observable but unstable (the z-domain poles are at $z = -0.6$ and -2).
   (ii) Bookwork.
   (iii) Reduced-order observer: state-space order is 2, and there is 1 output, hence we need to estimate 1 state. Choose $R = \begin{bmatrix} C \\ C_1 \end{bmatrix} = \begin{bmatrix} 1 & 0 \\ 0 & 1 \end{bmatrix}$. Therefore no similarity transformation is required, and we have

$$\begin{aligned} x_1(k+1) &= -x_1(k) + 0.8x_2(k) + u(k) = y(k+1) \\ x_2(k+1) &= 0.5x_1(k) - 1.6x_2(k) + 2u(k) \end{aligned}$$

Letting

$$\begin{aligned} \bar{q} &= y(k+1) + x_1 - u \\ \bar{u} &= 0.5x_1 + 2u \end{aligned}$$

gives the following reduced order state-space representation

$$\begin{aligned} x_2(k+1) &= -1.6x_2 + \bar{u} \\ \bar{q} &= 0.8x_2 \end{aligned}$$

The error dynamics between the state and its estimate is governed by this state-space's "A" and "C" matrices. That is

$$e(k+1) = [-1.6 - 0.8\ell]e(k)$$

The number $-1.6 - 0.8\ell$ defines the error's behaviour, and its magnitude needs to less than unity for stability. For rapid convergence of the error to zero, we set it to a small number, say 0.2 giving $\ell = -2.25$.
As discussed in the text we can eliminate $y(k+1)$ by defining $q = \hat{\bar{x}}_2 - Ly$. Then the complete state estimate is $\hat{\bar{x}} = \begin{bmatrix} \hat{\bar{x}}_1 \\ \hat{\bar{x}}_2 \end{bmatrix} = \begin{bmatrix} y \\ Ly + q \end{bmatrix}$. A block diagram of the reduced-order observer is shown in Figure D.2 (using the notation in chapter 4).

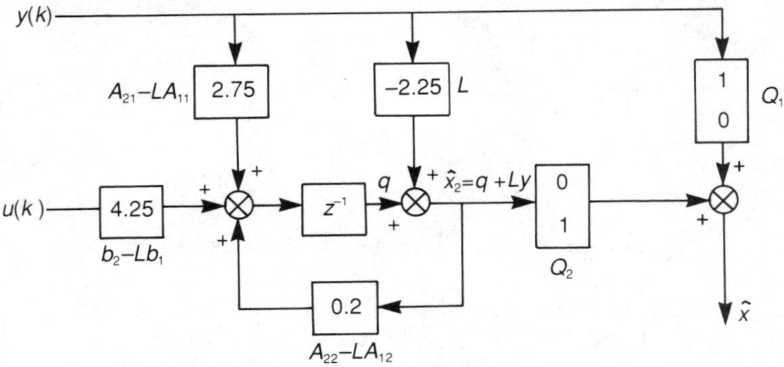

Figure D.2 *Reduced-order observer implementation*

The state estimate $\hat{x}$ can be multiplied by a time-varying gain calculated by the optimal control method and used as the state feedback signal.

8. It is advisable to draw the uncompensated root-locus to assess the performance and how improvements can be made. A rough sketch is shown in Figure D.3(a), from which it is clear that the system is sluggish and has very low damping. It is necessary to introduce extra poles and zeros to pull the locus into the unit circle. Points to bear in mind are
   - pole at $z = 1$ is an integrator, so system is type 1.

- settling time requirements dictate that the closed-loop poles must lie within a circle of radius $e^{-4.6T/t_s}$ for 1% error band. Since $T = 0.05$ s, and $t_s = 0.4$ s, this implies within a circle of radius 0.56.
- on z-domain root-locus we need to make the locus pass beyond the $\zeta = 0.7$ curve, and within the settling time constraint. Allowing a safety margin we will try to get the locus to pass $\approx 0.4 \pm j0.2$ (from z-plane root-locus paper).

Where does zero and pole of compensator need to be? Try putting zero on positive real axis and pole on negative real axis to pull locus in as required.

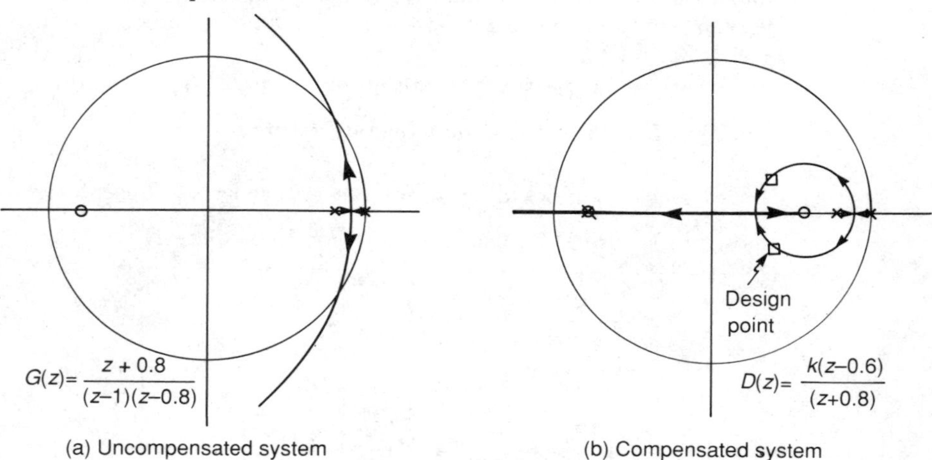

Figure D.3 z-plane root-locus design

Zero: needs to be approximately equal distance from the points $0.4 \pm j0.2$ and 0.9 (the breakaway point of the 2 system poles). By using simple trigonometry calculations, or measuring on the z-plane root-locus paper, we can determine that the zero needs to be placed at ≈ 0.6.

Pole: for simplicity put pole at $z = -0.8$ so that it cancels the system zero.

Therefore the proposed controller is

$$D(z) = K\frac{(z-0.6)}{(z+0.8)}$$

which gives rise to the root-locus shown in Figure D.3(b). The compensated system has then a closed-loop transfer function of

$$\frac{Y(z)}{U(z)} = K\frac{(z-0.6)}{(z-1)(z-0.8) + K(z-0.6)}$$

and a characteristic equation equal to

$$z^2 - (1.8 - K)\,z + 0.8 - 0.6K = 0$$

Therefore the closed-loop poles are at

$$\frac{1.8 - K \pm \sqrt{(1.8 - K)^2 - 4\,(0.8 - 0.6K)}}{2}$$

When the real part of this, that is, $(1.8 - K)/2 = 0.4$, we have $K = 1$, giving an imaginary part of $\pm j0.2$. Hence at $K = 1$ the closed-loop poles are at $0.4 \pm j0.2$ which are within specifications. Therefore the required controller is

$$D\,(z) = \frac{z - 0.6}{z + 0.8}$$

When a unit step is applied the output equals

$$Y\,(z) = z^{-1} + 1.2z^{-2} + 1.16z^{-3} + 1.09z^{-4} + \cdots$$

and the final value theorem gives $y_{ss} = 1$ as required.

9. (i) The continuous state-space is

$$\begin{bmatrix} \dot{x}_1\,(t) \\ \dot{x}_2\,(t) \\ \dot{x}_3\,(t) \end{bmatrix} = \begin{bmatrix} -1 & 0 & 0 \\ 0 & -3 & 0 \\ 0 & 0 & -5 \end{bmatrix} \begin{bmatrix} x_1 \\ x_2 \\ x_3 \end{bmatrix} + \begin{bmatrix} 3 & 0 \\ 2 & 1 \\ 0 & 4 \end{bmatrix} \begin{bmatrix} u_1 \\ u_2 \end{bmatrix}$$

$$\begin{bmatrix} y_1\,(t) \\ y_2\,(t) \end{bmatrix} = \begin{bmatrix} 3 & 0 & 0 \\ 0 & 2 & 5 \end{bmatrix} \begin{bmatrix} x_1\,(t) \\ x_2\,(t) \\ x_3\,(t) \end{bmatrix}$$

(ii) The discrete state-space form is:

$$\text{the system matrix} = \begin{bmatrix} 0.91 & 0 & 0 \\ 0 & 0.74 & 0 \\ 0 & 0 & 0.61 \end{bmatrix},$$

$$\text{and the input matrix} = \begin{bmatrix} 0.29 & 0 \\ 0.17 & 0.09 \\ 0 & 0.32 \end{bmatrix}.$$

(iii) For unit step inputs we have

$$y\,(0.1) = \begin{bmatrix} 0.78 \\ 1.33 \end{bmatrix}, \quad y\,(0.2) = \begin{bmatrix} 1.5 \\ 2.21 \end{bmatrix}, \quad y\,(0.3) = \begin{bmatrix} 2.16 \\ 2.76 \end{bmatrix}, \ldots$$

10. Clearly if a second-order transfer function is to be derived from a third-order state-space form, there is a pole–zero cancellation. Start from the state-space representation and convert to the transfer function form using

$$TF = C\,[zI - A]^{-1}\,B$$

11. The system is observable, uncontrollable and unstable (the poles are at $z = -0.5$ and -2). Only one mode is controllable, therefore if the unstable mode is uncontrollable, we cannot achieve objective. We can diagonalise the system to determine the uncontrollable mode. We need the eigenvector matrix, which can be shown to equal

$$M = \begin{bmatrix} 1 & 1 \\ -1 & 2 \end{bmatrix} \rightarrow M^{-1} = \frac{1}{3}\begin{bmatrix} 2 & -1 \\ 1 & 1 \end{bmatrix}$$

Performing a similarity transformation gives

$$M^{-1}AM = \begin{bmatrix} -0.5 & 0 \\ 0 & -2 \end{bmatrix}, \quad M^{-1}B = \begin{bmatrix} 0 \\ 1 \end{bmatrix}, \quad CM = \begin{bmatrix} 0.5 & 0.5 \end{bmatrix}$$

Hence the stable pole at -0.5 is uncontrollable, and the unstable pole at -2 is controllable. We will apply state feedback to the diagonal system. The compensated system matrix $A + BF = \begin{bmatrix} -0.5 & 0 \\ f_1 & -2 + f_2 \end{bmatrix}$ which yields a characteristic equation as

$$(z + 0.5)(z + 2 - f_2) = 0$$

and a closed-loop transfer function as

$$\frac{Y(z)}{U(z)} = \frac{0.5}{z + 2 - f_2}$$

When a unit input step is applied the final value theorem gives

$$y_{ss} = \lim_{z \to 1} \frac{z-1}{z} \frac{0.5}{z + 2 - f_2} \frac{z}{z-1}$$

For zero steady-state error $y_{ss} = 1$, therefore $f_2 = 2.5$, and f_1 can be anything (let it be equal to 0). The feedback required in terms of the original state-space form is given by

$$fM^{-1} = \frac{1}{3}[2.5 \quad 2.5]$$

If the states are inaccessible it is necessary to estimate them using an observer (full- or reduced-order). The system is observable so the whole state will be estimated. The observer dynamics are described by

$$\begin{aligned} \hat{x}(k+1) &= (A - LC)\hat{x}(k) + Ly(k) + Bu(k) \\ error(k+1) &= (A - LC)\,error(k) \end{aligned}$$

The characteristic equation of the error is

$$z^2 + (2.5 + 0.5\ell_1)z + 1 + 0.75\ell_1 - 0.25\ell_2$$

This must have faster eigenvalues than the system eigenvalues (which are at $z = 0.5$). We let the observer have 2 eigenvalues at 0.1. Therefore the required characteristic equation is $z^2 - 0.2z + 0.01 = 0$, and comparing coefficients gives the observer matrix $L = \begin{bmatrix} -5.4 \\ -12.2 \end{bmatrix}$.

12. System is controllable and observable, therefore we can estimate the state by using a state observer and use this to apply state feedback to allocate the pole locations to be at $0.4 \pm j0.3$. The separation principle means that the two operations can be done independently of each other.

First determine the state feedback vector $f = [f_1 \quad f_2]$. Here we have

$$A \rightarrow A + bf = \begin{bmatrix} f_1 & 1 + f_2 \\ -1 + 2f_1 & -2.5 + 2f_2 \end{bmatrix}$$

The new system matrix has a characteristic equation equal to

$$z^2 + (2.5 - 2f_2 - f_1)\, z + 1 - 4.5f_1 + f_2 = 0$$

The required characteristic equation is

$$z^2 - 0.8z + 0.25 = 0$$

Compering the two equations we can determine that $f = [0.47 \quad 1.41]$. An observer (reduced-order) can be constructed as follows: 2 states and 1 output, therefore we need to estimate 1 state. Using notation from chapter 4 we have $R = \begin{bmatrix} C \\ C_1 \end{bmatrix} = \begin{bmatrix} 1 & 1 \\ 0 & 1 \end{bmatrix}$. This gives $Q = R^{-1} = \begin{bmatrix} 1 & -1 \\ 0 & 1 \end{bmatrix} = \begin{bmatrix} Q_1 & Q_2 \end{bmatrix}$. Performing a similarity transformation on the system state-space, using R, gives a new state description as

$$\begin{aligned} \bar{x}\,(k+1) &= \begin{bmatrix} -2.5 & -0.5 \\ -3.5 & -1.5 \end{bmatrix} \bar{x}\,(k) + \begin{bmatrix} 3 \\ 2 \end{bmatrix} u\,(k) \\ y\,(k) &= \begin{bmatrix} 1 & 0 \end{bmatrix} \bar{x}\,(k) \end{aligned}$$

Partitioning the state into $\bar{x}_1$, which does not need estimation, and $\bar{x}_2$, which does, gives

$$\begin{aligned} \bar{x}_2\,(k+1) &= -1.5\bar{x}_2 + \bar{u} \\ \bar{q} &= 0.5\bar{x}_2 \end{aligned}$$

where $\bar{u} = -3.5\bar{x}_1 + 2u$, and $\bar{q} = y\,(k+1) - 2.5y - 3u$.
The observer equation is

$$\hat{\bar{x}}_2\,(k+1) = (-1.5 + 0.5\ell)\,\hat{\bar{x}}_2\,(k) + \ell\bar{q} + \bar{u}$$

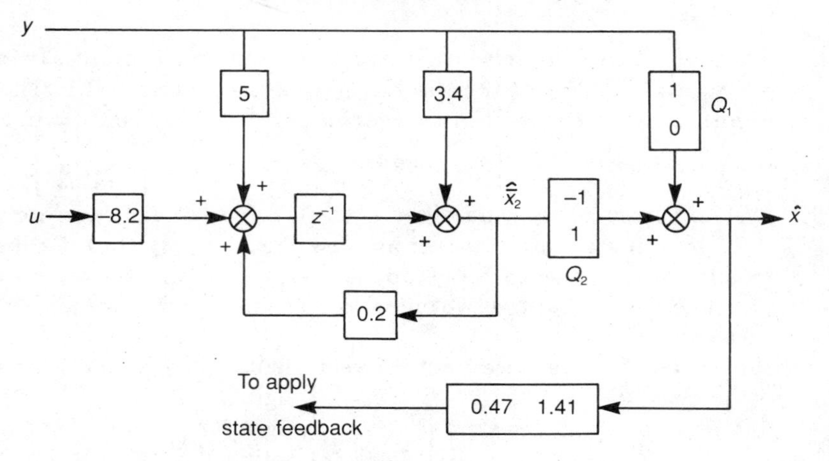

Figure D.4 State feedback and reduced-observer design

It is required that the dynamics of the observer should be faster than the closed-loop system (closer to the origin in the z-plane). Letting $(-1.5 + 0.5\ell) = 0.2$, gives $\ell = 3.4$. Then the state estimate can be generated and used in applying the state feedback design as shown in Figure D.4.

Index